SpringerBriefs in Electrical and Computer Engineering

T0075799

For further volumes:
www.springer.com/series/10059

Sid-Ahmed Selouani

Speech Processing
and Soft Computing

 Springer

Sid-Ahmed Selouani
Université de Moncton
Shippagan Campus
218, Boul. J-D Gauthier
Moncton
Canada
selouani@umcs.ca

ISSN 2191-8112   e-ISSN 2191-8120
ISBN 978-1-4419-9684-8   e-ISBN 978-1-4419-9685-5
DOI 10.1007/978-1-4419-9685-5
Springer New York Dordrecht Heidelberg London

Library of Congress Control Number: 2011936520

Printed on acid-free paper

Springer is part of Springer Science+Business Media (www.springer.com)

Preface

Soft Computing (SC) techniques have been recognized nowadays as attractive solutions for modeling highly nonlinear or partially defined complex systems and processes. These techniques resemble biological processes more closely than conventional (more formal) techniques. However, despite its increasing popularity, soft computing lacks a precise definition because it is continuously evolving by including new concepts and techniques. Generally speaking, SC techniques encompass two main concepts: approximate reasoning and function approximation and/or optimization. They constitute a powerful tool that can perfectly complement the well-established formal approaches when certain aspects of the problem to solve require dealing with uncertainty, approximation and partial truth. Many real-life problems related to sociology, economy, science and engineering can be solved most effectively by using SC techniques in combination with formal modeling. This book advocates the effectiveness of this combination in the field of speech technology which has provided systems that have become increasingly visible in a wide range of applications.

Speech is a very complex phenomenon involving biological information processing system that enables humans to accomplish very sophisticated communication tasks. These tasks use both logical and intuitive processing. Conventional 'hard computing' approaches have achieved prodigious progress, but their capabilities are still far behind that of human beings, particularly when called upon to cope with unexpected changes encountered in the real world.

Therefore, bridging the gap between the SC concepts and speech technology is the main purpose of this book. It aims at covering some important advantages that speech technology can draw from bio-inspired soft computing methods. Through practical cases, we will explore, dissect and examine how soft computing complement conventional techniques in speech enhancement and speech recognition in order to provide more robust systems.

This book is a result of my research, since 2000, at INRS-EMT Research Institute (Montreal, Canada) and LARIHS Laboratory in Moncton University (New Brunswick, Canada). Its content is structured so that principles and theory are often followed by applications and supplemented by experiments. My goal is to provide a cohesive vision on the effective use of soft computing methods in speech enhancement and speech recognition approaches.

The book is divided into two parts. Each part contains four chapters. Part I is entitled *Soft Computing and Speech Enhancement*. It looks at conventional techniques of speech enhancement and their evaluation methods, advocates the usefulness of hybridizing hierarchical connectionist structure with subspace decomposition methods, as well as the effectiveness of a new criterion to optimize the process of the subspace-based noise reduction. It also shows the relevance of evolutionary-based techniques in speech enhancement. Part II, *Soft Computing and Speech Recognition*, addresses the speech recognition robustness problem, and suggests ways that can make performance improvements in adverse conditions and unexpected speaker changes. Solutions involving Autoregressive Time-Delayed Neural Networks (AR-TDNN), genetic algorithms and Karhunen Loève transforms are explained and experimentally evaluated.

It is my hope that this contribution will both inspire and succeed in passing on to the reader my continued fascination with speech processing and soft computing.

Shippagan (NB), Canada   Sid-Ahmed Selouani

Contents

Chapter 1
Introduction

1.1 Soft Computing Paradigm

The concept of soft computing (SC) was introduced in the early nineties thanks to the pioneering ideas of Lotfi A. Zadeh [153]. Soft computing is inspired by the information processing in natural and biological systems that are capable to deal with uncertainty and imprecision to achieve robustness, tractability and optimal solutions. This leads to the emergence of computing approaches that provide the tolerability and stability when the systems are confronted with imprecise and/or distorted information. In contrast, hard computing, i.e., conventional computing, requires a precisely stated analytical model and is often valid under specific assumptions and for ideal cases.

The main constituents of soft computing are Fuzzy Logic (FL), Neurocomputing (NC), Evolutionary Computation (EC), Machine Learning (ML), Particle Swarm Optimization (PSO) and Probabilistic Reasoning (PR) which encompasses chaos theory and parts of learning theory. What is important to note is that these principal constituents are complementary rather than competitive. The combination of SC methods with conventional (more formal) approaches leads to the emerging field of hybrid intelligent systems that is in constant growth. In many real life applications, the advantages of building cooperative approaches using SC techniques at the same time as conventional modeling are demonstrated every day.

A glimpse into the future suggests that the role of soft computing will increase dramatically in the perspective of designing systems that are very close to human beings in their capabilities. In this context, speech processing constitutes a relevant field to experiment these new concepts, since the speech phenomenon reflects the remarkable ability of the human mind to process vagueness, ambiguity and imprecise information.

S.-A. Selouani, *Speech Processing and Soft Computing*, SpringerBriefs in Electrical and Computer Engineering, DOI 10.1007/978-1-4419-9685-5_1,
© Springer Science+Business Media, LLC 2011

1.2   Soft Computing in Speech Processing

The last few decades have seen massive advances in both speech technology and bio-inspired soft computing methods, but the two research fields have evolved by following parallel paths. The prodigious progress in calculation speed, parallel algorithms, and modeling open new perspectives towards the inclusion of more complexity; thus, more synergy between the two research fields is expected in the near future. Given that speech is primarily a biological signal and that soft computing techniques are generally inspired by biological processes, soft computing methods (in contrast to conventional techniques that are largely based on formal approaches) are better suited for tackling the many challenging problems of speech processing.

Neural networks and other SC approaches (mainly fuzzy logic and genetic algorithms) have proven effective in solving some pattern-matching problems in audio- and speech-processing applications. However, in contrast with other fields, very little research has been done to give soft computing the key role in such applications. Speech is a versatile, uncertain and imprecise phenomenon and remains in many situations (adverse conditions, unpredictable environments) intractable to conventional formalism and analytical methods. This book investigates the advantages that speech technology can draw from bio-inspired soft computing methods. It also attempts to provide new insights into speech recognition/enhancement obtained by investigating solutions beyond the statistical approach which dominates the field. The practical case studies presented aim at pointing to research directions on how conventional methods in speech processing and SC can benefit from each other.

1.3   Organization of the Book

The goal of this book is to permit both experienced researchers and graduate students to expand their horizon and to investigate new research directions through a review of the theoretical and practical settings of soft computing methods in very recent speech applications. In the first part of this book, we depict methods that use soft-computing to improve overall speech quality while minimizing any speech distortion. In the second part of this book, we deal with the application of SC in speech recognition and present some research directions on how these two fields can gain from each other.

The material discussed in the chapters of part I is as follows.

Chapter 2 reviews well-known methods of speech enhancement and provides an overview of the various assessment methods used to evaluate speech enhancement algorithms in terms of quality and intelligibility. Some conventional methods used for noise reduction are presented as a general introduction to the area of investigation. There are numerous ways in which speech enhancement methods can be classified. For the sake of simplification, we have divided these methods into four broad categories: spectral subtractive techniques, statistical-model-based techniques, subspace decomposition techniques and perceptual-based techniques. Common to these methods is their enhancement of speech degraded by an additive (background) noise where it is assumed that the speech signal is uncorrelated with the noise.

Chapter 3 focuses on signal (and feature) subspace filtering where a nonparametric linear estimate of the unknown clean-speech signal is obtained based on a decomposition of the observed noisy signal (or its representation into the feature space) into mutually orthogonal signal and noise subspaces. This decomposition is achievable by assuming a low-rank linear model for speech and an uncorrelated additive noise interference. We present signal and Mel-frequency subspace approaches that are based on the Karhunen-Love Transform (KLT) in a white noise degradation context. We also present a two-stage noise removal technique that combines Mel-frequency subspace decomposition and neural networks to provide more robust mel-frequency cepstral coefficients (MFCCs). The advantage of hybridizing SC-based (neural networks) and conventional enhancement techniques is then emphasized.

Chapter 4 introduces a method that optimizes the signal subspace decomposition on the basis of the variance or reconstruction error. This optimization is also used, later on in Chapter 8, in a hybrid scheme involving soft-computing techniques. Different methods for the selection of the number of principal components of the KLT are compared in a speech enhancement application using speech data in mobile communications.

Chapter 5 describes an approach based on a mapping process using a Mel-frequency subspace decomposition and genetic algorithms (GAs) to perform noise reduction. The goal of this evolutionary eigendomain KLT-based transformation is to achieve an enhancement of MFCCs under severe noise degradations. This chapter shows how the use of GAs overcomes the limits of estimating the noise variance and leads to the determination of robust solutions.

Part II of this book is dedicated to the use of soft-computing techniques in automatic speech recognition (ASR) systems. The content of the chapters of this part is summarized as follows.

Chapter 6 discusses the many challenges faced by speech technology in order to achieve efficient conversational systems. The ultimate goal consists of making ASR indistinguishable from the human understanding system. Here the problem of speech recognition robustness is analyzed. We present the three major approaches to cope with adverse conditions: signal compensation, feature space, and model adaptation techniques. In addition, the question of how to better represent the hearing/perception knowledge in ASR systems is covered in two sections since the extraction of reliable parameters is one of the most important issues in ASR. The relationship between ASR and human-system dialog is also explained. In the last section of this chapter, we present a new paradigm regarding the role of soft computing techniques in the robustness of speech recognition that may intervene at different processing levels: acoustic level, language level or understanding level.

Chapter 7 illustrates the usefulness of time-delayed recurrent neural networks incorporated in a hybrid structure through an approach which attempts to improve the performance of a modular ASR structure in the case of complex phonetic features. The objective here is to test the ability of a system combining the widely used Hidden Markov Models and neural networks to detect features as subtle as the nasalization feature of French vowels.

Chapter 8 presents a promising approach combining evolutionary-based subspace filtering to complement conventional ASR systems to tackle the challenge of noise robustness. In fact, the use of the soft computing technique based on GAs is less complex than many other robust techniques that need to either model or compensate for the noise. We also show that the knowledge gained from measuring the auditory physiological responses to speech stimuli may provide more robustness to speech recognition.

Chapter 9 presents a speaker adaptation technique characterized by a simplified adaptation process through the use of a single global transformation set of parameters optimized by genetic algorithms using a discriminative objective function. The goal here is to achieve adaptation, irrespective of the amount of available adaptive data. The framework presented in this chapter demonstrates the suitability of soft-computing technique (genetic algorithms) to improve unsupervised speaker adaptation using linear transforms.

1.4   Note to the Reader

This book is for any person who is interested in modern speech technology and/or in new applications of soft computing, particularly with techniques such as artificial neural networks, genetic algorithms, eigendecomposition, perception and hearing models. Research engineers as well as graduate students involved in speech and signal processing, telecommunications, human-to-computer interaction, artificial intelligence, pattern recognition, verbal interaction may also find this book very useful. The book aims at encouraging these potential readers to investigate speech enhancement and speech recognition through practical case studies that go beyond conventional approaches.

Part I
Soft Computing and Speech Enhancement

Chapter 2
Speech Enhancement Paradigm

Abstract Speech enhancement techniques aim at improving the quality and intelligibility of speech that has been degraded by noise. The goal of speech enhancement varies according to the needs of specific applications, such as to increase the overall speech quality or intelligibility, to reduce listener fatigue or to improve the global performance of an ASR embedded in a voice communication system. This chapter begins by giving a background on noise and its estimation and reviews some well-known methods of speech enhancement. It also provides an overview of the various assessment methods used to evaluate speech enhancement algorithms in terms of quality and intelligibility.

Keywords Speech enhancement • Noise • Spectral subtraction • Statistical techniques • Subspace decomposition • Perceptual methods • Enhancement evaluation

2.1 Speech Enhancement Usefulness

In many real-life contexts, there are a wide variety of situations in which we need to enhance speech signals. During the last few decades, the increasing use and development of digital communication systems has led to an increased interest in the role of speech enhancement in speech processing [14, 87, 36, 12, 88]. Speech enhancement techniques have been successfully applied to problems as diverse as correction of disrupted speech due to pathological problems of the speaker, pitch and rate modification, restoration of hyperbaric speech, and correction of reverberation, but noise reduction is probably the most important and most frequently studied issue. Voice communication over cellular telephone systems constitutes a typical environment where speech enhancement algorithms can be used to improve the quality of speech at the receiving back-end; that is, they can be used as a prepro-cessor in speech coding systems employed in mobile communication standards. In a device equipped with a speech recognition system, the speech quality and the

S.-A. Selouani, *Speech Processing and Soft Computing*, SpringerBriefs in Electrical and Computer Engineering, DOI 10.1007/978-1-4419-9685-5_2,
© Springer Science+Business Media, LLC 2011

automatic speech recognition performance may be degraded by the convolutional channel distortions and the additive environmental noises. In this case, the noisy speech signal can be preprocessed by a speech enhancement algorithm before being fed to the speech recognizer.

2.2 Noise Characteristics and Estimation

We are surrounded by noise everywhere, and these interfering sounds are present in different situations and forms in daily life. Prior to designing algorithms that cope with adverse conditions, it is crucial to understand the noise characteristics and the differences between the noise sources in terms of temporal and spectral characteristics. Noise can be impulsive, continuous, or periodic, and its amplitude may vary in frequency range.

2.2.1 Noise Characteristics

Noise can generally be classified into three major categories based on its characteristics:

- Stationary noise, i.e., remains unchanged over time, such as fan noise;
- Pseudo or Non-stationary noise, i.e., traffic or crowd of people speaking in the background, mixed in some cases with music;
- Transient noise, i.e., hammering or door slam.

The spectral and temporal characteristics of pseudo or non-stationary noise change constantly. Clearly, the task of suppressing this type of noise is more difficult than that of suppressing stationary noise. Another distinctive feature of noises is their spectrum shape, particularly the distribution of noise energy in the frequency domain. For instance, most of the energy of car noise is concentrated in the low frequencies, i.e., it is low-pass in nature. Train noise, on the other hand, is more broadband as it occupies a wider frequency range [2]. In most speech enhancement methods, the estimation of the power of the noise is a requirement. Fortunately, the bursty nature of speech makes it possible to estimate the noise during speech pauses. Moreover, it should be mentioned that it is easier to deal with additive noise than convolutive noise. This is why the assumption stating that the noise and speech are additive is often made.

For practical and natural reasons, the estimation of the noise is almost performed in the spectral domain. Actually, spectral components of speech and noise are partially uncorrelated. Besides this, perception/hearing and psycho-acoustic models are well understood (and adapted) in the spectral domain. Four domains of noise and noisy speech representation are used: spectral magnitude and power, the log-spectral power, the Mel-scale space for amplitude and power, and parametric such as autoregressive (AR) models. In each of these representation domains, knowledge of speech and noise intensity levels is critical in the design of most speech enhancement algorithms. Therefore, an estimate of the range of signal-to-noise ratio (SNR) level is frequently found in speech enhancement.

2.2.2 Noise Estimation

The most common model considers that noise is a Gaussian process with slow changes in its power spectrum. Noise and noisy speech spectrums are represented by spectral, mel-spectral or cepstral coefficients. A parametric representation could also be used as proposed by Ephraim *et al.* [40]. In this latter work, an all-pole spectrum model of the AR noise parameters provided by the Linear Predictive Coding (LPC) analysis is used as a robust front-end for a dynamic time warping based recognizer. Other studies present the noise model as a multi-state Hidden Markov Model (HMM) [54]. These noise estimation algorithms are based on statistical principles and are often used in speech recognition applications since they operate at the feature level (e.g., Mel-Frequency Cepstral Coefficients: MFCCs) in the log-spectral domain.

In most of real-life conditions, the spectral characteristics of the noise might be changing constantly. Therefore, there is a need to update the noise spectrum continuously over time, and this can be done by using noise-estimation algorithms. To face this challenge, one idea proposed by Rennie *et al.* [113] consists of modeling the noise as the sum of a slowly evolving component and a random component. Both components are represented in the Mel log power spectral domain. A first order Gaussian AR process in each frequency bin is used to model the evolving component, while the random component is considered as zero-mean Gaussian. This method provides update procedures for the mean and variance of the noise while the Gaussian mixture model for the speech remains fixed.

Several noise estimation techniques in frequently varying conditions have been proposed. The use of minimum statistics for noise estimation was introduced by Martin in [92]. It assumes that in any frequency bin there will be brief periods when the speech energy is close to zero and that the noise will then dominate the speech. Thus tracking the minimum power over a long frame makes it possible to estimate the noise level. Using averaging rather noise-only periods leads to a good estimates of noise. Another method proposed by Diethorn performs a moving average during noise-only durations [35]. The arithmetic average of the most recent noise-only frames of the signal is used as an estimate of the noise variance. Cohen proposed in [24] a minima controlled recursive algorithm (MCRA) which updates the noise estimate by tracking the noise-only periods by comparing the ratio of the noisy speech to the local minimum against a threshold. This method was improved by using a criterion applied on the noise-only regions of the spectrum and based on speech-presence probability [25].

To estimate the power of stationary noise, the first frames of noisy signals are usually assumed to be pure noise, and can therefore be used to estimate the noise. In the case of non stationary noise, the noise specifications need to be updated continuously. This can be done through a voice activity detector (VAD) that performs a speech/pause detection in order to update noise estimation. In the case of a non stationary noise or low SNR levels, the reliability of speech/pause detection is a concern. The next chapters will show that the use of soft computing in some speech enhancement frameworks may help us to avoid the explicit speech/pause detection (VAD) for noise estimation.

2.3 Overview of Speech Enhancement Methods

Numerous techniques have been proposed in the literature for speech enhancement. These techniques can roughly be divided into four main categories: spectral subtractive, statistical-model-based, subspace decomposition and perceptual based techniques.

2.3.1 Spectral Subtractive Techniques

Spectral subtraction was one of the earliest methods used for speech enhancement [14]. Spectral subtraction simply needs an estimate of the noise spectrum during periods of speaker silence (single channel) or from a reference source (multi-channel). It is a frame-based approach that estimates the short-term spectral magnitude of the noise-free signal from the noisy data. Spectral subtraction methods are based on the basic principle that as the noise is additive, one can estimate and update the noise spectrum when speech is absent and subtract it from the noisy signal spectrum to get an estimate of the clean signal spectrum. The main drawback of these methods is the introduction of an artificial noise called residual noise. Indeed, subtraction leaves peaks in the noise spectrum. On the one hand, the wider residual peaks result in broadband noise characterized by a time-varying energy. On the other hand, the narrower peaks that are separated by deep valleys in the spectrum, are the source of time varying tones termed as musical noise [88]. Different strategies have been proposed to increase subjective listening test quality and to reduce both distortion and musical noise [10, 138].

2.3.2 Statistical-model-based Techniques

Speech enhancement can be approached as a statistical estimation problem. The goal here is to find a linear (or non-linear) estimator of the original clean signal.

The Wiener and minimum mean-square error (MMSE) algorithms are among the well-known methods belonging to this category [38]. Three estimation rules known as maximum likelihood (ML), maximum *a posteriori* (MAP) and minimum mean-square error (MMSE) are known to have many desirable properties [88]. ML is often used for non-random parameters. MMSE estimation of speech signals, which have been corrupted by statistically independent additive noise, is optimal for a large class of difference distortion measures, provided that the posterior probability density function (PDF) of the clean signal given the noisy signal is symmetric about its mean. In addition to its optimality as pre-processor in autoregressive (AR) model vector Quantization (VQ) in the Itakura-Saito sense, the causal MMSE estimator is also the optimal pre-processor in minimum probability of error classification of any finite energy continuous time signal contaminated by white Gaussian noise [39]. However, the derivation of the MMSE estimator may be difficult, particularly in the case of complex statistical models for signal and noise. In this case, the maximum *a posteriori* (MAP) estimator using the expectation-maximization (EM) algorithm, can be useful. Wiener filters are considered as linear estimators of the clean speech signal spectrum and they are optimal in the mean-square sense. The enhanced time-domain signal is obtained by convolving the noisy signal with a linear (Wiener) filter. Equivalently, in the frequency domain, the enhanced spectrum is obtained by multiplying the input noisy spectrum by the Wiener filter.

2.3.3   *Subspace Decomposition Techniques*

These techniques are based on the principle that a nonparametric linear estimate of the unknown clean-speech signal is obtained by using a decomposition of the observed noisy signal into mutually orthogonal signal and noise subspaces. This decomposition is performed under the assumption that the energy of less correlated noise spreads over the entire observation space while the energy of the correlated speech components is concentrated in a subspace generated by the low-order components. The noise is assumed to be additive and uncorrelated with speech signal. Generally speaking, noise reduction is obtained by removing the noise subspace and by removing the noise contribution in the signal subspace [41]. The decomposition of the vector space of the noisy signal into subspaces can be done using the well-known orthogonal matrix factorization techniques from linear algebra namely, the Singular Value Decomposition (SVD) or the Eigen Value Decomposition (EVD). The idea to perform subspace-based signal estimation was originally put forward by Dendrinos *et al.* [33], who proposed the use of SVD on a data matrix containing time-domain amplitude values. Later on, Ephraim and Van Trees proposed a new technique based on EVD of the covariance matrix of the input speech vectors [41].

2.3.4 Perceptual-based Techniques

The goal of perceptual-based methods is to make the residual noise perceptually inaudible and therefore to improve the intelligibility of enhanced signals by considering the properties of the human auditory system. The idea is to exploit the fact that the hearing system cannot perceive residual noise when its level falls below the noise masking threshold (NMT). In these methods, the spectral estimates of a speech signal play a crucial role in determining the value of the noise masking threshold that is used to adapt the perceptual gain factor. Ching-Ta Lu in [22] proposed a two-step-decision-directed (TSDD) algorithm to improve the accuracy of estimated speech spectra. This spectral estimate is also used to compute the NMT which is applied to adapt a perceptual gain factor. This leads to a significant reduction of residual noise. A similar approach has been used by Ben Aicha and Ben Jebara [10] to eliminate only the perceptible parts of musical noise by using Wiener filtering and the detection of musical critical bands thanks to the tonality coefficient and a modified Johnston masking threshold. The results showed that a good trade-off was obtained between speech distortion and musical noise presence. We can divide perceptually-based methods into two sub-categories: methods incorporating auditory masking effects in the noise suppression rules implemented entirely in the perceptual domain, and methods employing a perceptual post-filter that exploits the masking properties to smooth the resulting enhanced speech.

2.4 Evaluation of Speech Enhancement Algorithms

Subjective listening tests involving the minimal pair phoneme contrasts remain the most accurate method for evaluating speech quality. The Mean Opinion Score (MOS) provides the most reliable method used by the subjective listening tests towards assessing speech quality. The principle consists of comparing the original and processed speech signals by involving a group of auditors who are asked to rate the quality of speech signal along a scale ranging from 1 to 5. Reliable subjective evaluation is costly and time consuming since it is conditioned by the choice of the listener panel and inclusion of anchor conditions. Therefore, many objective speech quality measures have been proposed to predict speech quality with high correlation with subjective speech quality measures such as MOS or Degradation MOS (DMOS) [106]. Ideally, an accurate objective speech quality measure would be able to assess the quality of enhanced speech without accessing the original speech. However, most of current objective measures are limited, since they require access to the original speech signal, and some can only model the low-level processing (e.g., masking effects) of the auditory system. Despite these limitations, some of these objective measures have been found to correlate well with subjective listening tests [88]. According to Yang et al. [149], these measures can be classified into three categories: time-domain measures, spectral domain measures and perceptual domain measures. In the following subsections, we briefly describe some these objective quality measures with respect to each category.

2.4.1 Time-Domain Measures

The simplest way to perform a time-domain measure consists of calculating the Signal-to-Noise Ratio (SNR) that performs a sample-by-sample comparison between original and processed speech signals. Speech waveforms are compared in the time domain. Therefore, the synchronization of the original and distorted speech is crucial. The most popular and accurate time-domain measure is the segmental signal-to-noise ratio (SegSNR). This measure is particularly effective in indicating the speech distortion than the overall SNR [141]. The frame-based segmental SNR is formed by averaging frame level SNR estimates. Higher values of the SegSNR indicates weaker speech distortions.

2.4.2 Spectral Domain Measures

These objective measures are generally calculated on speech frames that are typically between 15 and 30 ms long. The spectral measures are more accurate than the time-domain measures. They are also less sensitive to the time desynchronization between the original and the enhanced (or coded) speech [150]. Numerous spectral domain measures have been proposed in the literature including the log-likelihood ratio (LLR) measures [27], the cepstral distance measures [134], and the weighted slope spectral distance measure (WSS) [79]. The LLR measure is referred to as Itakura distance which compares the LPC vectors of the original signal with the LPC vectors of enhanced speech. Lower values of LLR measure indicate a better perceived quality. The cepstral measure is an euclidean distance between cepstral coefficients of the original signal and the processed signal. This distortion measure is considered as a human auditory measure and its higher values reflect important speech distortions. The WSS distance measure is based on an auditory model in which a number of overlapping filters of progressively larger bandwidth are used to estimate a weighted difference between the spectral slopes in each band. The magnitude of each weight indicates whether the band is near a spectral peak or valley, and whether the peak is the largest in the spectrum. The WSS measure is attractive because it does not require explicit formant extraction. A lower WSS reflects a better speech quality.

2.4.3 Perceptual Domain Measures

These measures perform a transformation of speech signals into perceptually relevant domains such as the bark spectrum and incorporate human auditory models. Among these measures, we can cite the Bark Spectral Distortion (BSD) measure developed at the University of California [141], the MBSD which is an improvement of BSD [150], and the perceptual evaluation of speech quality (PESQ) [69]. BSD was the first objective measure that incorporated psychoacoustic aspects. Its performance is considered better than conventional objective measures for speech coding distortions [120]. It assumes that the speech quality is linked to the speech loudness (magnitude of auditory sensation in psychoacoustics). This measure uses the Euclidean distance between loudness vectors of the original and enhanced speech to estimate the overall distortion. The difference between the BSD and the MBSD measure is that the latter includes a weighting function in the computation of the square difference of the frame-based loudness spectra. A global MBSD is computed by averaging the frames' values. Both the BSD and MBSD measures show a high correlation with the MOS score [150,88]. The lower the MBSD measure is, the better the quality of the enhanced signal is.

The PESQ is a widely-used and reliable method for assessing speech quality. The PESQ-based assessment, which is standardized in ITU-T recommendation P.862 [69], is performed by comparing a reference speech to the processed speech sample that has to be evaluated. Theoretically, the PESQ algorithm is designed to match the average of the listeners' opinion scores. PESQ provides a score ranging from 0.5 to 4.5. Higher scores indicate better quality. In the PESQ algorithm, the reference and the signal to evaluate (noisy) are level-equalized to a standard listening level. The gain of the two signals is not *a priori* known and may vary considerably. The gains of the reference, noisy and enhanced signals are calculated by using the root mean square values of filtered speech (350-3250 Hz). The signals are aligned in time to correct for time delays, and then processed through an auditory transform to obtain the loudness spectra. The difference, termed the disturbance, between the loudness spectra is computed and averaged over time and frequency to produce the prediction of subjective MOS score.

2.5 Summary

This chapter has provided an overview of speech enhancement algorithms. In each section we attempted to include descriptions of some of the well-known methods. This chapter also presented some procedures that have been used to evaluate the performance of speech enhancement algorithms. Enhancement algorithms can be evaluated in terms of speech intelligibility and speech quality. A description of common objective quality measures was also provided.

Chapter 3
Connectionist Subspace Decomposition for Speech Enhancement

Abstract In this chapter, a two-stage noise removal algorithm that deals with additive noise is proposed. In the first stage, a feedforward neural network (NN) with a backpropagation training algorithm is applied to match the uncorrupted information. In the second stage, the Karhunen-Loève Transform (KLT) based subspace filtering is used to compensate for the destruction caused by the noise. This combination is motivated by the fact that neural networks have the ability to learn from examples, even from complex relationships (non-linear) between inputs and outputs, and that subspace filtering has demonstrated its effectiveness to perform noise reduction through an optimal representation of features.

Keywords Subspace filtering • Eigenvalue decomposition • Singular value decomposition • Neural networks • KLT • Noise reduction

In the next sections, we describe the main subspace decomposition methods. We also present signal and mel-frequency subspace approaches based on the KLT in a white noise degradation context. The problem of the optimal component selection often encountered in subspace decomposition methods is also addressed. Then, the two-stage noise removal algorithm including the NN is explained and experimentally evaluated.

3.1 Method Overview

Signal subspace filtering is a class of speech enhancement techniques that has attracted much interest and inquiry. The principle of these methods is to determine a nonparametric linear estimate of the unknown clean speech signal by performing a decomposition of the observed noisy signal into mutually orthogonal signal and noise subspaces [62]. The basic implementation of these methods consists of

S.-A. Selouani, *Speech Processing and Soft Computing*, SpringerBriefs in Electrical and Computer Engineering, DOI 10.1007/978-1-4419-9685-5_3,
© Springer Science+Business Media, LLC 2011

applying the Principal Components Analysis (PCA), also known within the signal processing community as the Karhunen-Loève Transform (KLT), to the observed noisy signal. Traditional approaches for single-channel noise reduction use KLT to identify a set of orthogonal basis vectors that decompose the vector space of the noisy signal into a signal-plus-noise subspace and a noise subspace. The underlying assumption is that in the case of white noise, the energy of less correlated noise is distributed over the entire observation space isotropically while the energy of the correlated speech is almost concentrated in a small linear subspace. Therefore, the clean signal subspace can be estimated consistently by projecting the noisy observations onto the signal subspace. Noise reduction is obtained by discarding the rest of the samples that are generated by the remainder of the space orthogonal to the signal subspace. Thus, a certain amount of noise filtering is obtained by keeping only the components that are in the signal subspace defined by linear combinations of the first few most energized basis vectors.

3.2 Definitions

Principal components analysis (PCA) is known as a technique used to reduce multidimensional data to lower dimensions for analysis. It consists of computing the eigenvalue decomposition of a data set usually after mean centering the data for each attribute. PCA performs orthogonal transformation to convert a set of observations of possibly correlated variables into a set of uncorrelated variables called principal components. Each principal component has a higher variance than the succeeding one. Hence, the first component gives the direction where the maximum variance could be observed. PCA is also named the KarhunenLoève Transform. From the mathematical point of view, PCA performs an Eigen Value Decomposition (EVD) on the square matrix of covariance or correlation. In a more general case, a Singular Value Decomposition (SVD) is applied on a matrix of data which is not necessarily square. Mathematically EVD/PCA and SVD are equivalent. The only difference is that EVD/PCA is applied on the covariance (or correlation) matrices while SVD could be applied on any type of matrices. We use the KLT term to refer to the implementations of PCA/EVD method in speech enhancement and recognition.

3.3 Eigenvalue Decomposition

Let's consider that the stochastic processes $X(t)$, $N(t)$, and $S(t)$, that have generated $x(t)$, $n(t)$, and $s(t)$, respectively, are wide sense ergodic. We define a real-valued observation vector $x(t) \in \Re^K$ to be the sum of the signal vector $s(t) \in \Re^K$ and a noise vector $n(t) \in \Re^K$, i.e.,

$$x(t) = s(t) + n(t). \tag{3.1}$$

We arrange a K-dimensional observation vector in a $M \times N$ Hankel-structured observation matrix $\mathbf{X}_{M \times N}(t)$ where $K = M + N - 1$, i.e.,

$$\mathbf{X}_{M \times N}(t) = \begin{pmatrix} x_t & x_{t-1} & \cdots & x_{t-N+1} \\ x_{t-1} & x_{t-2} & \cdots & x_{t-N} \\ \vdots & \vdots & \ddots & \vdots \\ x_{t-M+1} & x_{t-M} & \cdots & x_{t-M-N+2} \end{pmatrix} \tag{3.2}$$

Due to the ergodicity assumption, we can estimate the correlation matrix R_{xx} using the zero-mean-scaled version of (3.2) as:

$$R_{xx} = \frac{1}{M-1} X^T X \in \Re^{N \times N}. \tag{3.3}$$

Let $q_1, q_2, ..., q_N$ be eigenvectors corresponding to the eigenvalues $\lambda_1, \lambda_2, ..., \lambda_N$ of the correlation matrix $R_{xx} \in \Re^{N \times N}$. By defining the $\mathbf{Q}$ matrix as:

$$\mathbf{Q} = [q_1 \, q_2 \, ... \, q_N,] \in \Re^{N \times N} \tag{3.4}$$

where the eigenvectors are orthonormal due to the symmetry in R_{xx} and the eigenvalues are ordered in decreasing order in a diagonal matrix:

$$\Lambda = diag(\lambda_1, \lambda_2, ..., \lambda_N) \in \Re^{N \times N}$$
$$\text{where } \lambda_1 \geq \lambda_2 \geq ... \geq \lambda_N \geq 0 \tag{3.5}$$

For positive-definite matrices, we can decompose the original matrix R_{xx} into its eigenvalue decomposition,

$$\mathbf{R}_{xx} = \mathbf{Q}\Lambda\mathbf{Q}^T. \tag{3.6}$$

Major signal subspace techniques assume the noise to be white with a σ_n^2 variance. Thus, the EVD of the noise autocorrelation matrix is given by:

$$\mathbf{R}_{nn} = \mathbf{Q}(\sigma_n^2 I)\mathbf{Q}^T, \tag{3.7}$$

where I is the identity matrix. Thus Equation 3.6 can be written as

$$\mathbf{R}_{xx} = \mathbf{Q}(\Lambda_{clean} + \sigma_n^2 I)\mathbf{Q}^T. \tag{3.8}$$

The diagonal matrix containing the eigenvalues of the clean signal is denoted by Λ_{clean}. The enhancement of corrupted speech is performed by assuming that the clean component is concentrated in an $r < N$ dimensional subspace (signal subspace) while the noise occupies the $N-r$ dimensional space. Therefore the noise reduction is performed by considering only the signal subspace. This is obtained by nullifying the components in the noise subspace. This operation needs prior knowledge of the signal dimension to correctly define the signal subspace, which we called the optimal order of reconstruction. Numerous approaches for estimating the order of a good reconstruction model are found in the literature [62].

3.4 Singular Value Decomposition

The Singular Value Decomposition (SVD) gives relevant information about the structure of a matrix. Its main advantage is that it works on both square and rectangular matrices. In practice, the SVD constitutes a useful numerical tool to perform noise reduction since it applies directly on the observation vector. Thanks to SVD, any rectangular data matrix can be transformed into a diagonal matrix resulting in the following matrix factorization:

$$\mathbf{X} = \mathbf{U}\boldsymbol{\Sigma}\mathbf{V}^T, \tag{3.9}$$

where $\mathbf{U}$ and $\mathbf{V}$ are the singular orthonormal matrices defined by:

$$U = [u_1, ..., u_M] \in \Re^{M \times M}, \tag{3.10}$$

and

$$V = [v_1, ..., v_N] \in \Re^{N \times N}. \tag{3.11}$$

The vectors u_i and v_i are called the *ith* left singular vector and the *ith* right singular vector of $\mathbf{X}$, respectively. $\boldsymbol{\Sigma}$ is a diagonal matrix whose entries are the singular values that are always positive:

$$\boldsymbol{\Sigma} = diag[\sigma_1, \sigma_2..., \sigma_p] \in \Re^{M \times N}, p = min(M, N). \tag{3.12}$$

The SVD can be used to improve the SNR by setting the noise related singular values equal to zero. To obtain this enhancement, one approach consists of modifying the structure of $\mathbf{U}$, $\mathbf{V}$ and $\boldsymbol{\Sigma}$ matrices by truncating them from rows and columns that contribute to the noise. This process requires the determination of the optimal rank K which will permit one to determine the clean estimate $\tilde{\mathbf{X}}$,

$$\tilde{\mathbf{X}} = \tilde{\mathbf{U}}\tilde{\boldsymbol{\Sigma}}\tilde{\mathbf{V}}^T, \tag{3.13}$$

where $\tilde{\mathbf{U}}$ and $\tilde{\mathbf{V}}^T$ are $\Re^{N \times K}$ and $\Re^{M \times K}$ respectively.

The new estimated matrix $\tilde{\mathbf{X}}$ has the characteristic that its rank is exactly the same as a dimension of singular matrices. The optimality of this new clean estimate is reached by setting the $|\tilde{\mathbf{X}} - \mathbf{X}|^2$ distance to satisfy the minimality condition. However, due to the impact of noise on the signal related singular values, $\tilde{\mathbf{X}}$ loses its Hankel structure. Thanks to the method initially proposed by Cadzow [17], it is possible to restore the Hankel structure of $\hat{\mathbf{X}}$, while keeping the rank at K and to reach iteratively the estimate optimality. The Hankel structure can be recovered according to the least squares criterion. Let's define $\mathbf{H}$ as a related matrix with Hankel structure. The criterion consists of minimizing the sum $\sum |\tilde{x}_{ij} - h_{nm}|^2$. Differentiating to the $\mathbf{H}$ matrix elements and using the Hankel property $h_{nm} = h_{n+m}$ yields

$$\sum 2(\tilde{x}_{ij} - d_{n+m}h_{n+m}) = 0, \tag{3.14}$$

which can be written as follows:

$$h_{n+m} = \sum \frac{\tilde{x}_{ij}}{d_{n+m}}, \tag{3.15}$$

where d_{n+m} is the number of matrix elements on the $n + m'^{th}$ anti-diagonal of $\mathbf{H}$. The Cadzow theory constitutes the basis for all subspace-based enhancement methods. It is worth noting that these techniques share one key step towards noise reduction, which is to precisely determine the limits of the clean subspace into which the projection will be performed. The advantage of working with the SVD is that no explicit estimation of the covariance matrix is needed. Nevertheless, all estimators can be performed by using the EVD-based scheme.

3.5   KLT Model Identification in the Mel-scaled Cepstrum

One of the key issues in using subspace decomposition methods is the selection of the number of principal components (PCs) for the clean speech estimate. There are many methods for calculating the number of optimal PCs, but most of them use monotonically increasing or decreasing indices which makes the decision to choose the number of principal components very subjective. If fewer PCs are selected than required, a poor model will be obtained, which results in an incomplete representation of the process. Conversely, if more PCs than necessary are retained, the model will be over-parameterized and will include noise. Different approaches have been developed to select the optimal number of PCs: (1) Akaike Information Criterion (AIC) [6], (2) Minimum Description Length (MDL) [116], (3) Cumulative Percent Variance (CPV) [89], (4) Average Eigenvalue (AE) [86], (5) Parallel Analysis (PA) [82], and (6) Variance of the Reconstruction Error (VRE) [105, 2].

Instead of dealing with the speech signal, we chose to use the noisy Mel-Frequency Cepstral Coefficients (MFCC) vector. The cepstral coefficients are used to describe the short-term spectral envelope of a speech signal. The cepstrum is the inverse Fourier transform of the logarithm of the short-term power spectrum of the signal. By means of the logarithmic operation, the vocal tract transfer function and the voice source are separated. The advantage of using such coefficients is that they reduce the dimension of a speech spectral vector while maintaining its identity. There are two ways to obtain the cepstral coefficients: FFT cepstral and LPC cepstral coefficients. In the derivation of cepstral coefficients, the Mel-scale is currently widely used because this scale improves the performance of speech recognition systems over the traditional linear scale. More details on the calculation of these derived MFCC are given in Section 6.3.1.

A KLT is performed on the noisy zero-mean normalized MFCC vector $\hat{\mathbf{C}} = [\hat{C}_1, \hat{C}_2, ..., \hat{C}_N]^T$. Assuming that $\hat{\mathbf{C}}$ has a symmetric non-negative autocorrelation matrix $R = \mathcal{E}[\hat{\mathbf{C}}^T \hat{\mathbf{C}}]$ with a rank $r \leq N$, $\hat{\mathbf{C}}$ can be represented as a linear combination of the eigenvectors $\beta_1, \beta_2, ..., \beta_r$, that correspond to the eigenvalues $\lambda_1 \geq \lambda_2 \geq .... \geq \lambda_r \geq 0$, respectively. That is, $\hat{\mathbf{C}}$ can be calculated using the following orthogonal transformation:

$$\hat{\mathbf{C}} = \sum_{k=1}^{r} a_k \beta_{\mathbf{k}}, \quad k = 1, ..., r, \tag{3.16}$$

where the coefficients a_k, which are called the principal components, are given by the projection of the vector $\hat{\mathbf{C}}$ in the space generated by the eigenvector basis, as follows:

$$a_k = \hat{\mathbf{C}}^T \beta_{\mathbf{k}}, \quad k = 1, ..., r. \tag{3.17}$$

In [41], the linear estimation of the clean vector $\mathbf{C}$ is performed using two perceptually meaningful estimation criteria as follows:

$$\tilde{\mathbf{C}} = \sum_{k=1}^{r} W_k a_k \beta_{\mathbf{k}}, \quad k = 1, ..., r, \tag{3.18}$$

where W_k is a weighting function given by:

$$W_k = \left[\frac{\lambda_k}{\lambda_k + \sigma_n^2} \right]^{\gamma}, \quad k = 1, ..., r, \tag{3.19}$$

where σ_n^2 is the noise variance and $\gamma \geq 1$ (to be fixed experimentally).

An alternative choice for W_k which results in a more aggressive noise suppression is given by:

$$W_k = \exp\left\{ \frac{-\nu \sigma_n^2}{\lambda_k} \right\}, \quad k = 1, ..., r. \tag{3.20}$$

The value of the parameter ν is to be fixed experimentally.

Speech enhancement is performed by removing the noise subspace and estimating the clean signal from the remaining signal space. This estimation is done by projecting the noisy vectors in the subspace generated by the low-order components, given the fact that the high-order eigenvalues are more sensitive to noise than the low-order ones.

Since the weighting functions are used in the cepstral domain, we propose to deal with not only one of the time-dedicated weighting functions proposed by Ephraim [41], given by Equations 3.19 and 3.20 but with a combination of them. That is, for the component of the cepstrum which is less corrupted by noise, the weights g_i defined by Equation 3.19 are used. On the other hand, the weights defined by Equation 3.20 are used for highly corrupted signals. In fact, this was not chosen in an arbitrary way, but our choice was guided by what is called the *reconstruction's quality function*, denoted by Q. We defined Q as the ratio of the sum of the eigenvalues used to reconstruct the MFCC vector, to the sum of all the eigenvalues, as follows:

$$Q = \frac{\sum_{k=1}^{r} \lambda_k}{\sum_{k=1}^{N} \lambda_k}. \tag{3.21}$$

The first- and second-order derivatives of Q are given by:

$$\Delta Q = \frac{\lambda_{r+1}}{\sum_{k=1}^{N} \lambda_k} \tag{3.22}$$

and

$$\Delta \Delta Q = \frac{\lambda_{r+1} - \lambda_r}{\sum_{k=1}^{N} \lambda_k}$$

$$= \frac{\lambda_{r+1} - \lambda_r}{N\sigma_n^2 + \sum_{k=1}^{N} \lambda_k^{clean}}. \tag{3.23}$$

where λ_k^{clean}, $k = 1, ..., N$ are the eigenvalues of the clean signal. Given the fact that the magnitudes of the low-order eigenvalues are higher than the magnitudes of the high-order ones, the effect of the noise on the low-order eigenvalues is less than that of high-order ones. Thus, the variations of $\Delta \Delta Q$ for a certain noise variance σ_n^2 and a certain value of r tend to zero for higher order eigenvalues. Consequently, the Q-acceleration function ($\Delta \Delta Q$) helps us determining the optimal component order, denoted r_{th} at which we switch between the use of the two weighting functions defined by Equations 3.19 and 3.20.

3.6 Two-Stage Noise Removal Technique

The two-stage noise removal algorithm consists of two levels of MFCC enhancement. As illustrated in Figure 3.1, at the first level, the noisy 13-dimensional vector (12 MFCCs + energy) is fed to a Multi-Layer Perceptron in order to reduce the noise effects on such a vector. This first pre-processing does not require any *a priori* knowledge about the nature of the corrupting noise which theoretically permits

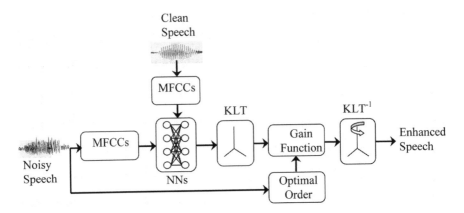

Fig. 3.1 The Hybrid MLP-KLT-based speech enhancement system.

dealing with any kind of noise. Moreover, this approach avoids the noise estimation process requiring a speech/non-speech pre-classification, which could be inaccurate for low SNRs. However, the MLP training requires a large amount of data [53]. Once we obtain the enhanced vector, it is fed to the second stage where a KLT is performed. This represents the second enhancement level that aims at refining the enhanced vector by projecting it into the subspace generated by optimized weighted eigenvectors. The motivation behind the use of a second level of enhancement after using the MLP network is to compensate for the limited power of the MLP network for enhancement outside the training space [53]. The fact that the noise and the speech signal are combined in a nonlinear way in the cepstral domain justifies the use of MLP, since it can approximate the required nonlinear function to some extent [58, 57]. The input of the MLP is the noisy MFCC vector $\mathbf{C}'$, while the actual response of the network $\hat{\mathbf{C}}$ is computed during a training phase using a convergence algorithm to update the weight vector in a manner that minimizes the error between the output $\hat{\mathbf{C}}$ and the desired clean cepstrum value $\mathbf{C}$. The weights of this network are calculated during a training phase with a back-propagation training algorithm using a mean square error criterion [99].

3.7 Experiments

In the following experiments the TIMIT database, described in [133], was used. The TIMIT corpus contains broadband recordings of a total of 6300 sentences, 10 sentences spoken by each of the 630 speakers from 8 major dialect regions of the United States, each reading 10 phonetically rich sentences. To simulate a noisy environment, car noise was added artificially to the clean speech. The MLP network was trained using noisy speech at different values of SNR varying from

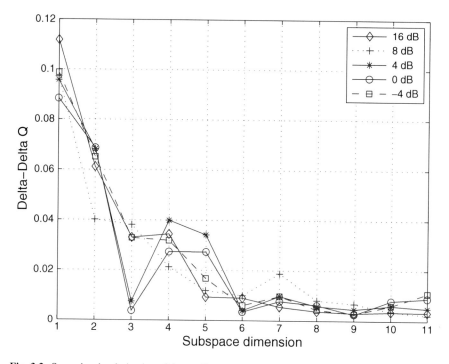

Fig. 3.2 Second order derivative of the quality reconstruction function Q.

16 dB to −4 dB. The architecture of the network that has been used throughout all experiments consists of three layers. The input layer is composed of 13 neurons, while the hidden layer and the output layer are composed of 26 and 13 neurons, respectively. The input to the network is the noisy 12-dimensional MFCC vector in addition to the energy. The weights of this network are calculated during a training phase with a back-propagation algorithm with a learning rate equal to 0.25 and a momentum coefficient equal to 0.09. The eigenvectors used in the KLT reconstruction module are weighted by the gain function according to the optimal choice of W_k given by either Equation 3.19 or 3.20. In our experiments, the optimal choice was based on the variations of $\Delta\Delta Q$. These variations are shown in Figure 3.2 for different SNR values. We found through experiments as shown in Figure 3.2 that $r = 6$ is a convenient value for switching between the use of either Equation 3.19 or 3.20 for the computation of the reconstructed vector given by Equation 3.18. It was found that the use of such a combination leads to an optimization of such weights. Figure 3.3 shows the first four MFCCs for a signal that has been chosen from the test set of the TIMIT corpus. It is clear from the comparison illustrated in this figure that the processed MFCCs, using the two-stage approach, are less variant than the noisy MFCCs and closer to the original ones.

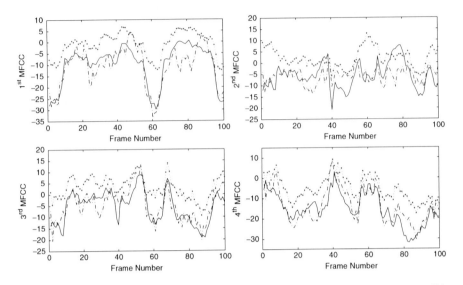

Fig. 3.3 Comparison between clean, noisy and NN-KLT-enhanced MFCCs represented by solid, dashed and dot-dash lines respectively.

3.8 Summary

In this chapter a connectionist subspace decomposition enhancement method operating in the mel-scaled cepstrum was presented. The noise reduction is performed by a two-stage system. In the first stage, a NN-based procedure was applied to 'learn' about the noise and to bring the noisy speech closer to the clean speech. At the same time the uncorrupted information is kept as well as possible. In the second stage, a KLT-based enhancement is performed by projecting the noisy signal into a subspace generetad by optimal noise-free components. The enhanced MFCCs parameters are found very close to the clean ones.

Chapter 4
Variance of the Reconstruction Error Technique

Abstract A critical issue in developing a KLT-based speech enhancement model is to select the optimal number of principal components (PCs). If fewer PCs are considered by the model, relevant components of speech may be lost. Conversely, if more PCs are selected, the model will be ineffective and the noise will remain. The purpose of this chapter is to present a signal subspace decomposition method using a Variance of Reconstruction Error (VRE) criterion to optimally select principal components. A benchmarking of various methods for selecting the number of PCs in a speech enhancement application is performed using data from the NOIZEUS database.

Keywords PCA model • Reconstruction error • Minimum Description Length • Spectral Subtraction • Objective measures • NOIZEUS database

4.1 General Principle

During the reconstruction process using the PCA model, the reconstruction error is a function of the number of PCs. Qin and Dunia in [105] use the variance of the reconstruction error (VRE) to determine the number of principal components. The VRE is decomposed into the principal component subspace and a residual subspace. The portion in the principal component subspace has a tendency to increase with the number of PCs, and that in the residual subspace has a tendency to decrease, resulting in a minimum in VRE. In the case of noisy component identification and reconstruction, the VREs are weighted based on the variance of each variable. The VRE based selection criterion has the advantage to work with both correlation-based and covariance-based PCA.

As proposed by Abolhassani *et al.* [2], the VRE method is used to determine the number of PCs based on the best reconstruction of the estimate of clean speech. A prominent point about this approach is that the proposed index has a minimum (i.e., non-monotonic) corresponding to the best reconstruction.

S.-A. Selouani, *Speech Processing and Soft Computing*, SpringerBriefs in Electrical and Computer Engineering, DOI 10.1007/978-1-4419-9685-5_4,
© Springer Science+Business Media, LLC 2011

4.2   KLT Speech Enhancement using VRE Criterion

Generally speaking, the variable reconstruction using PCA consists of estimating one variable from the others by exploiting the redundancy between these variables. Therefore, the accuracy of the reconstruction after projecting the noisy signal into the assumed noise-free subspace is related to the PCA capacity of revealing the redundancy among variables, which is closely related to the number of components. Let's assume that our signal is corrupted with a noise n_j along a direction $\xi_j \in \Re^N$:

$$x = s + n_j \xi_j \tag{4.1}$$

where $\|\xi_j\| = 1$.

The task of signal reconstruction is to find an estimate for s along the direction ξ_j. In other words, we correct the signal along the noise direction such that:

$$\hat{s} = x - \hat{n}_j \xi_j \tag{4.2}$$

has minimum model error, i.e.,

$$n_j = \arg\min_{n_j} \|s - \hat{s}\| = \arg\min_{n_j} \|\tilde{s}\|^2 = \arg\min_{n_j} \|\tilde{x} - \tilde{n}_j \tilde{\xi}_j\|^2 \tag{4.3}$$

where $\tilde{s}$ and $\hat{s}$ show the clean signal portion in the residual subspace and principal component subspace respectively.

The *argmin* of Equation 4.3 can easily be found through the use of least squares,

$$\hat{n}_j = \frac{\hat{\xi}_j^T \tilde{x}}{\hat{\xi}_j^T \hat{\xi}_j} = \frac{\hat{\xi}_j^T x}{\hat{\xi}_j^T \hat{\xi}_j} \tag{4.4}$$

Substituting the above $\hat{n}_j$ into Equation 4.2 we obtain the best signal reconstruction as follows.

$$\hat{s} = \left(I - \frac{\tilde{\xi}_j \tilde{\xi}_j^T}{\tilde{\xi}_j^T \tilde{\xi}_j} \right) x = \left(-\tilde{\xi}_j^o \tilde{\xi}_j^{oT} \right) x \tag{4.5}$$

where $\tilde{\xi}_j^o \equiv \tilde{\xi}_j / \|\tilde{\xi}_j\|$. using Equations 4.1 and 4.2, we can write,

$$s - \hat{s} = (\hat{n}_j - n_j)\xi_j \tag{4.6}$$

Substituting Equation 4.1 into Equation 4.4 leads to

$$\hat{n}_j - n_j = \frac{\tilde{\xi}_j^T s}{\tilde{\xi}_j^T \tilde{\xi}_j}. \tag{4.7}$$

Thus, the reconstruction error is given by:

$$s - \hat{s} = (\hat{n}_j - n_j)\xi_j = \frac{\tilde{\xi}_j^T s}{\tilde{\xi}_j^T \tilde{\xi}_j}\xi_j \tag{4.8}$$

and

$$\|s - \hat{s}\| = \frac{|\tilde{\xi}_j^T s|}{\tilde{\xi}_j^T \tilde{\xi}_j}. \tag{4.9}$$

From Equations 4.8 and 4.9 we can note that the variance of $(s - \hat{s})$ occurs only in the reconstruction direction ξ_j (same as for the variance of $n_j - \hat{n}_j$). The reconstruction error $(s - \hat{s})$ depends on the number of PCs retained in the PCA model. The number of optimal PCs is then obtained by achieving the minimum reconstruction error. Interestingly, if the reconstruction error is minimized for a particular magnitude n_j, it is minimized for all magnitudes. Therefore, it becomes possible to determine the number of PCs for the case of $n_j = 0$ to achieve the best reconstruction. Assuming $n_j = 0$, the variance of the reconstruction error in the direction ξ_j can be calculated as follows:

$$u_j \equiv \text{var}\{\xi_j^T(x - \hat{s})\} = \text{var}\{\hat{n}_j\} = \frac{\tilde{\xi}_j^T R_{xx}\tilde{\xi}_j}{\left(\tilde{\xi}_j^T \tilde{\xi}_j\right)^2} \tag{4.10}$$

where u_j is the variance of the reconstruction error in the estimation of s by using $\hat{s}$ and R_{xx} is the correlation matrix defined in Equation 3.3. In order to find the number of PCs, the u_j has to be minimized with respect to the number of PCs. Considering different noise directions and summing u_j in all dimensions, the VRE to be minimized is given by

$$VRE(l) = \sum_{j=1}^{N} \frac{u_j(l)}{\xi_j^T R_{xx}\xi_j}. \tag{4.11}$$

4.2.1 Optimized VRE

In order to equalize the importance of each variable, variance-based weighting factors may be applied. The VRE algorithm used in conjunction with EVD or SVD can be summarized by Algorithm 4.1. In order to make the determination of the optimal order of reconstruction more accurate, the minimum over the D previous frames can be considered,

$$l_{opt} = \min\left(\arg\min_l [VRE_{t-D}(l)]\right), \tag{4.12}$$

Data: noisy speech frames
Result: EVD/SVD optimal order for noise reduction
while *not at end of number of lags (D)* **do**
    | calculate the u_j and VRE using Equations 4.10 and 4.11;
    | determine the minimum of VRE ;
    | **if** $u_j >= var(\xi_j^T x)$ **then**
    | | put $\hat{s} = 0$ in Equation 4.10;
    | **else**
    | | increment the number of lags;
    | **end**
end
calculate l_{opt} according to Equation 4.12;

Algorithm 4.1: VRE Algorithm

---

where t is the frame index and D is the number of past frames (Lags) that are used to determine the optimal number of components. Equation 4.12 gives more robustness to the process of determining the order of reconstruction and prevents the rapid fluctuations of that optimal order that could be due to artefacts. The suggested value of D is 3.

4.2.2 Signal Reconstruction

To reconstruct a signal from noisy observations, the noise-only subspace should be removed and the remaining signal subspace should be modified to eliminate the effect of noise from this subspace. Ephraim and Trees proposed in [41] two estimates of the clean signal: the Time Domain Constrained (TDC) and the Spectral Domain Constrained (SDC). In the TDC estimator, the signal distortion is minimized while the residual noise energy is maintained below a user-defined upper bound thanks to a control parameter. The SDC estimate consists of minimizing the signal distortion for a fixed spectrum of the residual noise. The speech signal masks the residual noise and results in a filter having a gain function which is solely dependent on the desired spectrum of the residual noise. In the experiments presented in this chapter, the variant of the TDC estimate is used. The TDC estimator described in [114] is given by

$$\hat{S} = X\hat{Q}G_\mu \hat{Q}^T, \tag{4.13}$$

where $\hat{Q}$ is the truncated matrix of eigenvectors obtained by removing the last $N - l_{opt}$ columns of the original Q. The optimal rank l_{opt} is obtained by the procedure of Algorithm 4.1. In [41], only white noise is considered, thus G_μ is a diagonal matrix containing l_{opt} diagonal

$$g_\mu(m) = \frac{\lambda_{clean}(m)}{\lambda_{clean}(m) + \mu\sigma_w^2},\tag{4.14}$$

where σ_w^2 is the variance of the white noise and $\lambda_{clean}(m)$ is the clean signal variance in the mth dimension, and μ is the Lagrange multiplier. After estimating $\hat{\mathbf{S}}$ using the modified TDC estimator, the clean signal is estimated by averaging the antidiagonal values of $\hat{\mathbf{S}}$.

4.3 Evaluation of the KLT-VRE Enhancement Method

4.3.1 Speech Material

To evaluate the performance of the VRE-based enhancement technique, extensive objective quality tests are carried out with the NOIZEUS database [67]. NOIZEUS contains 30 IEEE sentences spoken by three male and three female speakers, corrupted by eight different real-world noises added artificially at different SNRs taken from the AURORA database [64]. The thirty sentences were selected so as to include all phonemes of American English. These sentences were initially sampled at 25 kHz and downsampled to 8 kHz. The frame sizes are 30 ms long with 40% overlap and a Hamming window is used. As detailed in [67], to simulate the telephone handset characteristics, both speech and noise signals were filtered by the filters used in the ITU-T P.862 [69] standard using the PESQ measure. In these experiments, four different noisy conditions are included: babble (crowd of people), car, exhibition hall, and train are considered.

4.3.2 Baseline Systems and Comparison Results

The methods to be compared with VRE are the Minimum Description Length (MDL) [116], Wiener and the Spectral Subtraction (SS) methods [88]. The objective measures used for the evaluation are the Weighted Spectral Slope (WSS) distance (smaller reflects less distortion) and the Perceptual Evaluation of Speech Quality (PESQ). These measures are chosen because they are strongly related to the subjective intelligibility that is correlated to speech recognition performance. In VRE and MDL, $N = 21$ (KLT dimension). In Wiener, α (the smoothing factor for the Decision-Directed method for estimation of a priori SNR) equals 0.99, and the smoothing factor for the noise updating is 9. In SS, c (the scaling factor in silence periods) is set to 0.03. Table 4.1 gives a comparison of the performance of the different methods including VRE. In this table, WSS and PESQ achievements of the enhanced signals as well as the SNRs of the original noisy signals are shown.

Table 4.1 Objective WSS and PESQ evaluations of subspace decomposition techniques using MDL and VRE criteria compared with SS and Wiener baseline techniques. NOIZEUS database is used for the evaluation. Best scores are highlighted in boldface.

Objective Measures	Input SNR (db)	Wiener	SS	MDL	VRE
WSS	0	156.82	178.42	184.73	**156.76**
	5	112.35	122.34	128.32	**108.74**
	10	**75.21**	85.96	89.84	75.96
PESQ	0	1.38	1.15	1.16	**1.42**
	5	1.94	1.92	1.98	**2.62**
	10	2.69	2.88	2.58	**2.95**

[a] Objective evaluation under NOIZEUS babble noise degradation.

WSS	0	122.34	161.54	158.74	**121.58**
	5	**92.26**	100.98	101.68	93.28
	10	**61.72**	78.05	81.62	69.34
PESQ	0	**1.52**	1.38	1.07	1.50
	5	2.88	1.89	1.25	**2.98**
	10	**2.84**	1.95	1.64	2.79

[b] Objective evaluation under NOIZEUS car noise degradation.

WSS	0	166.34	159.12	171.65	**152.96**
	5	**126.58**	127.25	120.14	127.52
	10	**92.57**	85.48	92.76	97.84
PESQ	0	2.40	2.56	1.98	**2.48**
	5	3.09	3.12	2.88	**3.19**
	10	3.52	3.48	3.04	**3.56**

[c] Objective evaluation under NOIZEUS exhibition hall noise degradation.

WSS	0	138.24	148.78	140.42	**135.54**
	5	102.47	103.57	103.98	**102.18**
	10	**67.98**	66.38	72.96	68.21
PESQ	0	1.78	1.69	1.45	**1.88**
	5	2.04	**2.28**	2.12	2.20
	10	**3.75**	3.48	3.15	3.72

[d] Objective evaluation under NOIZEUS train noise degradation.

Each table represents different noisy conditions. The results show that in most of the noisy situations (except in the case of the car noise where the Wiener seems to be slightly better) the VRE technique performs better than other methods. The comparison of the time-domain signals of the original clean and noisy signals as well as the output of the KLT-VRE given in Figure 4.1 shows the good performance of the enhancement method. This illustration is carried out on the sentence *The speaker announced the winner* uttered by a male speaker and corrupted by white Gaussian noise at an input SNR = 1 dB.

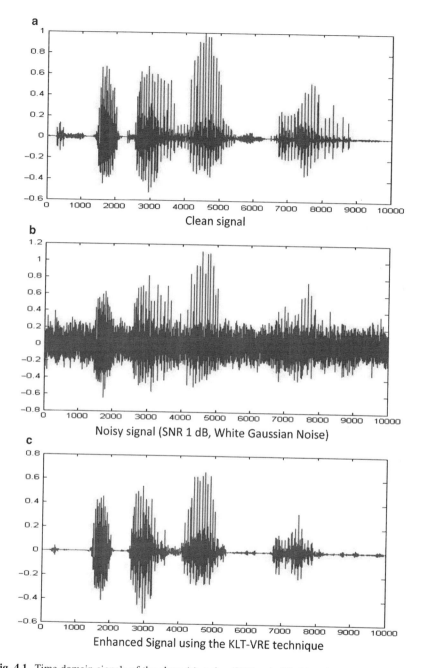

Fig. 4.1 Time-domain signals of the clean (a), noisy (SNR = 1 dB) (b) and KLT-VRE enhanced signals. The sentence *The speaker announced the winner* is uttered by a male speaker and corrupted by a white Gaussian noise

4.4 Summary

In this chapter a promising subspace approach for speech enhancement in noisy environments is presented. This approach is based on principal component analysis and optimal subspace selection using the variance of reconstruction error. The performance evaluation based on objective measures show that the VRE-based approach achieves a lower signal distortion and a higher noise reduction than existing enhancement methods. A prominent point of this subspace method is that it can be used as noise-robust front-end processing for speech recognizers, as we will see in Chapter 8.

Chapter 5
Evolutionary Techniques for Speech Enhancement

Abstract Genetic Algorithms have become increasingly appreciated as an easy-to-use general method for a wide range of optimization problems. Their principle consists of maintaining and manipulating a population of solutions and implementing a 'survival of the fittest' strategy in their search for better solutions. In this chapter, GAs are combined with a signal subspace decomposition technique to enhance speech that is severely degraded by noise. To evaluate the effectiveness of this hybrid approach, a set of continuous speech recognition experiments is carried out by using the NTIMIT telephone speech database.

Keywords Genetic Algorithms • KLT • Mel-frequency cepstral coefficients • Telephone speech • Channel degradations • NTIMIT database

5.1 Principle of the Method

Genetic algorithms (GAs) are a subset of evolutionary computation [37] that mimic the process of natural evolution. To perform such process, GAs implement mechanisms inspired by biological evolution such as selection, recombination and mutation, applied in a pool of individuals belonging to a same population. The fittest individuals, that represent parameters to optimize, are encouraged to reproduce and survive to the next generation, thus improving successive generations. A small proportion of inferior individuals can also be selected to survive and also reproduce. Recombination and mutation create the necessary diversity and therefore facilitate novelty, while selection is used to increase quality. Many aspects of such an evolutionary process are stochastic. In this chapter, GAs are used to overcome the limit of estimating the noise variance in subspace methods. The idea is to exploit the power of GAs to investigate beyond the classical space of solutions by exploring a wide range of promising areas [131, 123]. The approach consists of combining subspace decomposition methods and GAs as a means to determine robust solutions.

S.-A. Selouani, *Speech Processing and Soft Computing*, SpringerBriefs in Electrical and Computer Engineering, DOI 10.1007/978-1-4419-9685-5_5,
© Springer Science+Business Media, LLC 2011

5.2   Global Framework of Evolutionary Subspace Filtering Method

The principle of subspace decomposition methods consists of constructing an orthonormal set of axes that point in the directions of maximum variance and the enhancement is performed by estimating the noise variance. As described in the previous chapters, the enhancement is performed by assuming that the clean speech is concentrated in an $r < N$ dimensional subspace (signal subspace) whereas the noise occupies the $N - r$ dimensional observation space. In their pioneering work, Ephraim and Van Trees [41], the noise reduction is obtained through an optimal estimator that would minimize the speech distortion considering the fact that the residual noise fell below a preset threshold. The determination of such a threshold requires a noise variance estimation.

Mathematically, the subspace filtering consists of finding a linear estimate of s (the clean signal) given by $\hat{s} = Hx$, which can be written $\hat{s} = Hs + Hn$ where H is the enhancement filter and x the noisy signal. The filter matrix H can be written as: $H = QGQ^T$ in which the diagonal matrix G contains the weighting factors g_i for the eigenvalues of the noisy speech. In the evolutionary eigendecomposition, the H matrix becomes H_{gen} and is given by the following: $H_{gen} = QG_{gen}Q^T$ in which the diagonal matrix G_{gen} contains weighting factors that are optimized using genetic operators. Optimization is reached when the Euclidean distance between C_{gen} and C, the genetically enhanced and original parameters respectively, is minimized. The space of feature representation is reconstructed by using the eigenvectors weighted by the optimal factors of the G_{gen} matrix.

By using GAs, no empirical or *a priori* knowledge is needed. The problem of determining optimal order of reconstruction r is avoided since the GA implicitly discovers this optimal order. The complete space dimension N is considered at the beginning of the evolution process. As illustrated in Figure 5.1, the space of feature representation is reconstructed by using the eigenvectors weighed by the optimal factors of G_{gen} matrix.

5.3   Hybrid KLT-GA Enhancement

The evolution process starts with the creation of a population of the weight factors, g_i, which constitute the individuals. The individuals evolve through many generations in a pool where genetic operators are applied [49]. Some of these individuals are selected to reproduce according to their performance. The individuals' evaluation is performed through the use of an objective function. When the fittest individual (best set of weights) is obtained, it is used, in the test phase, to project the noisy data. Genetically modified MFCCs, their first and second derivatives, are finally used as enhanced features.

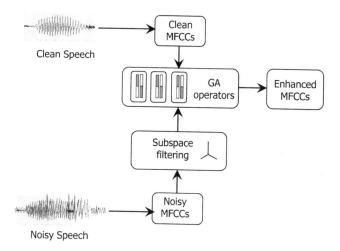

Fig. 5.1 General overview of the KLT-GA-based system.

5.3.1 Solution Representation

A solution representation is needed to describe each individual g_i in the population. A useful representation of individuals involves genes or variables from an alphabet of floating point numbers with values varying within upper and lower bounds a_i, b_i respectively. Concerning the initialization of the pool, the ideal zero-knowledge assumption is to start with a population of completely random values of weights. These values follow a uniform distribution within the upper and lower boundaries.

5.3.2 Selection Function

Selection is the process of determining the number of trials a particular individual is chosen for reproduction. The selection method used here, is the Stochastic Universal Sampling (SUS) introduced by Baker [8]. It consists of transforming raw fitness values into a real-valued expectation of an individual's probability to reproduce, and then to perform the selection based on the relative fitness of individuals. To do that, the individuals g_i are mapped to contiguous segments of a line such that the length of each individual's segment is equal to the value of its fitness. Equal space pointers are then placed over the line as many as the predetermined number (N_s) of individuals to select. The complete SUS procedure is given by Algorithm 5.1.

Data: population of g_k, and N_s
Result: index k of individuals selected to reproduce
order population by fitness;
Calculate F_t the total fitness of the population ;
Determine a random number $Rand$ between 0 and F_t/N_s;
for $i \leftarrow 0$ **to** $N_s - 1$ **do**
    calculate $f = rand + i * F_t/N_s$;
    ptr=0;
    while *not at end of population* **do**
        if $ptr < f$ *and fitness of* $g_k + ptr > f$ **then**
            return k;
        end
        ptr=ptr+fitness of fitness of g_k;
    end
end

Algorithm 5.1: Stochastic universal sampling algorithm for individual selection

5.3.3 Crossover and Mutation

To avoid the extension of the exploration domain in order to reach the best solution, a simple crossover operator can be used [65]. It generates a random number l from a uniform distribution and undergoes an exchange of the genes of the parents (X and Y) on the offspring genes (X' and Y'). It can be expressed by the following equations:

$$\begin{cases} X' = lX + (1-l)Y \\ Y' = (1-l)X + lY. \end{cases} \tag{5.1}$$

In addition to the crossover operator, a mutation is performed. Mutation consists of altering one or more gene values of the individual. This can result in entirely new individual. Through this manipulation, the genetic algorithm prevents the population from stagnating at a given non optimal solution. Usually the mutation rate is low (as in the biological world) and it is fixed according to a user-definable value. If this value is set very high, the search will become a random search. Most mutation methods in canonical GAs are randomly driven. Some methods such as that proposed by Temby *et al.* [132] suggest the use of directed mutation based on the concept of the momentum commonly used in the training of neural networks.

The principle of the mutation-with-momentum algorithm used here, requires that a gene's value has both the standard Gaussian mutation and a proportion of the current momentum term added to it. The update of the momentum term is performed to reflect the combined mutation value. The following equations summarizes the process of Gaussian mutation with momentum. Some individuals are selected and then their original genes, x, produce mutant genes, x_m,

$$\begin{cases} x_m = x + \mathcal{N}(0,1) + \eta M_0 \\ M_m = x_m - x \end{cases} \tag{5.2}$$

where $\mathcal{N}(0, 1)$ is a random variable of normal distribution with zero mean and standard deviation 1 which is to be sampled for each component individually, and η is the parameter controlling the amount of momentum ($0 < \eta < 1$). M_0 is the value of the momentum term for the gene and M_m is the value of the momentum term after mutation. The momentum is updated at each iteration by substituting M_0 by M_m. To prevent the momentum term from becoming large, the difference of Equation 5.2 is limited to a maximum value of M_m.

5.4 Objective Function and Termination

Evolution is driven by an objective function defined in terms of a distance measure between the noisy MFCCs, projected by using the individuals (weights), and the clean MFCCs. The fittest individual is the set of weights which corresponds to the minimum of that distance. As we are using MFCCs, Euclidean distance is considered. The GA must search all the axes generated by the KLT of the Mel-frequency space to find the closest to those of the clean MFCCs. The fittest individual is the axis corresponding to the minimum of that distance. Let's consider two vectors $\mathbf{C}$ and $\hat{\mathbf{C}}$ representing two frames, each with N components, where the geometric distance is defined as:

$$d(\mathbf{C}, \hat{\mathbf{C}}) = \left(\sum_{k=1}^{N} (\mathbf{C_k} - \hat{\mathbf{C}_k})^l \right)^{1/l}. \tag{5.3}$$

The Euclidean distance corresponds to ($l = 2$). The opposite of this distance, $-d(\mathbf{C}, \hat{\mathbf{C}})$ is used since we have to maximize the fitness function. The evolution process is terminated when a certain number of maximum generations is reached. This number corresponds to the beginning of the objective function convergence.

5.5 Experiments

The evaluation of the hybrid KLT-GA enhancement method is carried out by testing its robustness as it performs a speech recognition over a telephone channel. It is well-known that the limitation of the analysis bandwidth in the telephone channel yields higher speech recognition error rates. In these experiments a HMM-based speech recognition system is trained with high-quality speech and tested by using simulated telephone speech.

Fig. 5.2 Model of the telephone channel [77].

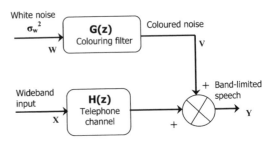

5.5.1 Speech Databases

In the first set of experiments, the training set providing the clean speech models is composed of the Train subdirectories of the TIMIT database described in [133]. The speech recognition system uses the Test subdirectories of NTIMIT as a test set [72]. The NTIMIT database was created by transmitting TIMIT sentences over a physical telephone network. Previous work on speech recognition systems has demonstrated that the use of speech over the telephone line yields a reduction in accuracy of about 10% [96]. The model used to simulate the telephone channel is described in [77]. Figure 5.2 shows that the wideband input sequence corresponding to TIMIT speech, is bandlimited by $H(z)$, the transfer function simulating the frequency response characteristics of a telephone channel. The channel noise is created by passing zero mean white noise with variance through a second filter $G(z)$ producing a coloured noise. This coloured noise is added to the $H(z)$ output to obtain the telephone speech. In the second set of experiments, NTIMIT is used for both training and test.

5.5.2 Experimental Setup

The baseline HMM-based speech recognition system is designed through the use of the HTK toolkit [66]. Here three systems are compared: the KLT-based system as detailed in Chapter 3, the KLT-GA-based ASR system and the baseline HMM-based system which uses MFCCs and their first and second derivatives as input features (MFCC_D_A). The parameters used to control the run of the genetic algorithm are as follows. The initial population is composed of 250 individuals and was created by duplicating (cloning) the elements of the weighting matrix. In order to insure convergence, we allow the population to evolve through 300 generations, even if no improvement in fitness is observed beyond 200 generations, as is shown in Figure 5.3. The percentages of crossover rate and mutation rate are fixed respectively at 35% and 3%. The number of total runs was fixed at 80. In order to make an adequate comparison with the baseline front-end, after the GA processing, the MFCCs static vectors are expanded to produce a 39-dimensional (static+dynamic) vector.

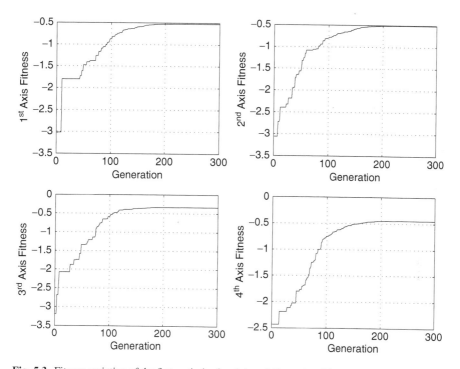

Fig. 5.3 Fitness variation of the first optimized weights of **G** matrix with respect to the number of generations within the evolutionary process.

5.5.3 Performance Evaluation

The results presented in Table 5.1 show that the use of the KLT-GA as a pre-processing approach to enhance the MFCCs that were used for recognition with 8-mixture Gaussian HMMs using tri-phone models, leads to a significant improvement in the accuracy of speech recognition. A correct rate of 34.49% is reached by the KLT-GA-MFCC_D_A-based CSR system when the baseline and the KLT-baseline (using the MDL criterion) systems achieve 18.02% and 27.73% respectively. This represents an improvement of more than 16% compared to the baseline system. Expanding to more than 8 mixtures did not improve the performance. Another set of experiments is carried out by applying the Cepstral Mean Normalization (CMN) to the MFCCs prior to the evolutionary subspace filtering. CMN is a widely used method for improving the robustness of speech recognition to channel distortions. The principle of CMN consists of performing a bias subtraction from the observation sequence (MFCC vector) resulting in a sequence that has a zero mean vector. In these experiments, the CMN included in the HTK toolkit is used [66]. The results show that the CMN has a significant impact on the baseline system using the MFCCs and their derivatives. An improvement of more than 4% is noticed.

Table 5.1 Percentages of word recognition rate ($\%C_{Wrd}$), insertion rate ($\%\epsilon_{Ins}$), deletion rate ($\%\epsilon_{Del}$), and substitution rate ($\%\epsilon_{Sub}$) of the MFCC_D_A, KLT-MFCC_D_A, and KLT-GA-MFCC_D_A ASR systems using 8-mixture tri-phone models. The cepstral mean normalisation (CMN) is also tested as preprocessing for all ASR systems. (Best rates are highlighted in boldface).

	$\%\epsilon_{Sub}$	$\%\epsilon_{Del}$	$\%\epsilon_{Ins}$	$\%C_{Wrd}$
MFCC_D_A	78.02	3.96	40.83	18.02
KLT-MFCC_D_A	66.95	5.32	31.74	27.73
KLT-GA-MFCC_D_A	60.85	4.66	29.56	**34.49**
[a] TIMIT is used for the training and NTIMIT for the test.				
MFCC_D_A	47.02	2.78	24.62	50.20
KLT-MFCC_D_A	39.36	3.37	18.62	57.61
KLT-GA-MFCC_D_A	33.64	3.24	10.39	**63.12**
[b] NTIMIT is used for the training and NTIMIT for the test.				
CMN-MFCC_D_A	74.54	3.28	38.36	22.18
CMN-KLT-MFCC_D_A	66.09	5.76	30.48	28.15
CMN-KLT-GA-MFCC_D_A	60.39	4.79	29.85	**34.82**
[c] TIMIT is used for the training and NTIMIT for the test. The CMN preprocessing is applied to the MFCCs.				
CMN-MFCC_D_A	43.58	2.04	21.58	54.38
CMN-KLT-MFCC_D_A	40.30	3.28	18.47	56.42
CMN-KLT-GA-MFCC_D_A	33.42	3.35	10.56	**63.23**
[d] NTIMIT is used for the training and NTIMIT for the test. The CMN preprocessing is applied to the MFCCs.				

However, the effect of the CMN preprocessing is very limited on the systems using KLT and GAs. Indeed, for the CMN-KLT-GA-MFCC_D_A, a little decrease (less than 1%) is noticed when NTIMIT is used for both training and test.

5.6   Summary

The approach described in this chapter can be viewed as a transformation via a mapping operator using a Mel-frequency subspace decomposition and GAs. The results show that this evolutionary eigendomain KLT-based transformation achieves an enhancement of MFCCs in the context of telephone speech. The improvement obtained over telephone lines demonstrates that the KLT-GA hybrid enhancement scheme succeeds in obtaining less-variant MFCC parameters under telephone-channel degradation. This indicates that both subspace filtering and GA-based optimization gained from the hybridization of the two approaches. It should be noted that the use of soft-computing technique leads to less complexity than many other enhancement techniques that need to either model or compensate for the noise.

Part II
Soft Computing and Automatic Speech Recognition

Chapter 6
Robustness of Automatic Speech Recognition

Abstract Most speech recognition research has shifted to conversational and natural speech in order to make effective and intuitive speech-enabled interfaces. Despite significant advances, many challenges remain to achieve the realization of efficient conversational systems. The ultimate goal consists of making ASR indistinguishable from the human understanding system. This chapter addresses the ASR robustness problem. It begins by giving the statistical formalism of speech recognition and then it describes the main robust features representing the hearing/perception knowledge that are used in the subsequent chapters. The major approaches used to achieve noise robustness are described. The relationship between dialog management systems and ASR is also investigated. Finally, a new paradigm giving soft computing techniques a new role in speech interactive systems is presented.

Keywords Speech recognition • Robustness • Mel-frequency cepstral coefficients • Auditory model • Acoustic indicative features • Dialog management

6.1 Evolution of Speech Recognition Systems

Automatic speech recognition has made enormous progress over the last two decades. Advances in both computing devices and algorithm development have facilitated these historical changes. In general, ASR can be viewed as successive transformations of the acoustic micro-structure of the speech signal into its implicit phonetic macro-structure. The main objective of any ASR system is to achieve the mapping between these two structures. To reach this goal, it is necessary to suitably describe the phonetic macro-structure, which is usually hidden behind the general knowledge of phonetic science [5, 36]. Thus, according to Deng [34], it is necessary to unify acoustic processing and to adapt the architecture of the ASR system to cover the broadest range of languages and situations. Modern configurations for ASR are mostly software architectures that generate a sequence of word hypotheses validated by a language model from an acoustic signal. The most popular and effective

S.-A. Selouani, *Speech Processing and Soft Computing*, SpringerBriefs in Electrical and Computer Engineering, DOI 10.1007/978-1-4419-9685-5_6, © Springer Science+Business Media, LLC 2011

algorithm implemented in these architectures is based on Hidden Markov Models (HMMs), which belong to the class of statistical methods [73]. Other approaches have been developed, but due to the complexity of their usability, they are still considered as research and development tools. Among these techniques, we can cite the one using hybrid neural networks and HMMs [15].

Speech recognition technology is already capable of delivering a good performance for many practical applications. For instance, we can already see in certain cars, interfaces that use speech recognition in order to control useful functions. Speech recognition is also used for telecommunication directory assistance. For Personal Computer (PC) users, Microsoft offers an improved speech recognition engine in its Windows Vista and Windows 7 operating systems [16]. It is now possible to control the system using speech commands. However, its biggest problem is that users need to learn a fixed set of commands. Mac OSx also provides a basic speech recognition engine. Nevertheless, to use it, users must create macros for each program, which can become quite tedious. The challenges of ASR are related to the use of robust acoustic features and models in noisy and changing environments; the use of multiple word pronunciations and efficient constraints allowing one to deal with a very large vocabulary and a variety of accents; the use of multiple and exhaustive language models capable of representing various types of situations and contexts; the use of complex methods for extracting conceptual representations and various types of semantic and pragmatic knowledge from pronounced utterances. The problem of ASR robustness and efficiency of spoken dialog systems is addressed in the following sections.

6.2   Speech Recognition Problem

ASR methods build speech sound models based on large speech corpora that attempt to include in their construction some common sources of variability that may occur in practice. Nevertheless, not all variability can reasonably be covered. For this reason, the performance of current ASR systems whose designs are predicated on predetermined conditions, degrades rapidly in the presence of adverse and/or unexpected conditions. Thus, the aim of a robust speech recognition system is to compensate for any type of mismatched conditions. In order to cope with mismatched (adverse) conditions and to achieve more robustness, numerous approaches have been proposed [13].

The ASR process aims at giving the ability to recognize speech sounds by comparing their acoustic features with those determined during the training. Thus, speech recognition is a pattern classification issue. From a probabilistic perspective, the ASR can be formulated using the Bayesian statistical framework. Let w refer to a sequence of phones or words, which produces a sequence of observable acoustic data o, sent through a noisy transmission channel. The recognition process aims to provide the most likely phone sequence w' given the acoustic data o. This estimation is performed by maximizing the *a posteriori* (MAP) $p(w/o)$ probability:

$$w' = argmax_{w \in \Psi} p(w/o) = argmax_{w \in \Psi} p(o/w)p(w) \tag{6.1}$$

where Ψ is the set of all possible phone sequences, $p(w)$ is the prior probability determined by the language model that the speaker utters w, and $p(w/o)$ is the conditional probability that the acoustic channel produces the sequence o. Let Λ be the set of models used by the recognizer to decode acoustic parameters through the use of the MAP procedure. Then Equation 6.1 can be written as follows:

$$w' = argmax_{w \in \Psi} p(w/o, \Lambda)p(w) \tag{6.2}$$

The mismatch between the training and testing environments induces a corresponding mismatch in the likelihood of o given Λ and consequently involves a breakdown of ASR systems. Decreasing this mismatch should increase the correct recognition rate.

HMMs constitute the most successful approach developed for modeling the statistical variations of speech in an ASR system. Each individual phone (or word) is represented by an HMM. In large-vocabulary recognition systems, HMMs usually represent subword units, either context-independent or context-dependent, to limit the amount of training data and storage required for modeling words. Most recognizers use typically left-to-right HMMs, which consist of an arbitrary number of states N. The observation sequence $\mathbf{O}$ possibly of different lengths, is assumed to be representative of the utterance to be recognized. The probability of the input observation vector is represented by the most common choice of distributions, the multivariate mixture Gaussian:

$$b_j(\mathbf{O}_t) = \sum_{m=1}^{M} c_{jm} \mathcal{N}(\mathbf{O}_t; \mu_{jm}, \Sigma_{jm}) \tag{6.3}$$

where M is the number of mixture components, c_{jm} is the weight of each mixture component of state j in each mixture and $\mathcal{N}(\mathbf{O}; \mu, \Sigma)$ denotes a multivariate Gaussian of mean μ and covariance Σ and can be written as:

$$\mathcal{N}(\mathbf{O}; \mu, \Sigma) = \frac{1}{\sqrt{(2\pi)^n |\Sigma|}} \exp^{-\frac{1}{2}(\mathbf{O}-\mu)' \Sigma^{-1}(\mathbf{O}-\mu)} \tag{6.4}$$

The performance of any recognition system depends on many factors, but the size and the perplexity of the vocabulary are among the most critical ones. A language model (LM) is essential for effective speech recognition. Typically, the LM will restrict the allowed sequences of words in an utterance. It can be expressed by the formula giving the *a priori* probability, $p(w)$

$$p(w) = p(w_1, ...w_m) = p(w_1 \prod_{i=2}^{m} p(w_i \mid \underbrace{wi - n + 1, ...w_{i-1}}_{n-1})) \tag{6.5}$$

In the n-gram approach, n is typically restricted to $n = 2$ (bigram) or $n = 3$ (trigram).

6.3   Robust Representation of Speech Signals

ASR systems use parameters to represent the waveform of a speech utterance. The extraction of reliable parameters is one of the most important issues in ASR. There are a large number of features that can be used for ASR. This parameterization process serves to maintain the relevant part of the information within a speech signal while eliminating the irrelevant part for the ASR process. A wide range of possibilities exists for parametrically representing the speech signal such as: short-time spectral envelope, Linear Predictive Coding (LPC) coefficients, MFCCs, short-time energy, zero crossing rates and other related parameters. It has been shown through several studies that the use of human hearing properties provides insight into defining a potentially useful front-end speech representation [36]. The digital filter bank method is one of the algorithms based on auditory functions and often used in speech ASR front-ends. A filter bank can be regarded as a model of the initial transformation in the human auditory system. Three choices for the frequency axis of this bank of filters could be used in such analysis: uniform spacing (as in the standard FFT), exponential spacing (a Constant-Q or wavelet transform) or perceptually-derived spacing. A mapping of the acoustic frequency to a perceptual frequency scale could be defined in the *bark* scale or *mel* scale [102]. Mel-scale filter banks are used to compute mel-frequency cepstral coefficients that have been shown to be favorable in ASR and are widely used in many ASR systems. Beside the filter-bank-based techniques, perceptual properties have also been integrated into the analysis of the speech signal through other algorithms such as the *Perceptual Linear Predictive (PLP)* analysis [60] and the so-called *relative spectra* (RASTA) techniques [61]. Including these auditory-based pre-processing techniques in ASR systems has led to an improvement of their performances. However, the performance of current ASR systems is far from the performance achieved by humans. The following subsections depict two methods that represent the hearing/perception knowledge in ASR systems.

6.3.1   Cepstral Acoustic Features

Among all parameterization methods, the cepstrum has been shown to be favorable in ASR and is widely used in many ASR systems [36] [102]. The cepstrum is defined as the inverse Fourier transform of the logarithm of the short-term power spectrum of the signal. The use of a logarithmic function permits us to deconvolve the vocal tract transfer function and the voice source. Consequently, the pulse sequence originating from the periodic voice source reappears in the cepstrum as a strong peak in the 'quefrency' domain. The derived cepstral coefficients are commonly used to describe the short-term spectral envelope of a speech signal. The advantage of using such coefficients is that they induce a data compression of each speech spectral vector while maintaining the pertinent information it contains. Davis and

Mermelstein in [28] introduced the use of the Mel-scale in the derivation of cepstral coefficients. The Mel-scale is a mapping from a linear to a nonlinear frequency scale based on human auditory perception. An approximation to the Mel-scale is:

$$Mel(f) = 2595 log_{10}\left(1 + \frac{f}{700}\right), \tag{6.6}$$

where f corresponds to the linear frequency scale. It is proved that such a scale significantly increases the performance of speech recognition systems in comparison with the traditional linear scale. The computation of MFCCs requires the selection of M critical bandpass filters. To obtain the MFCCs, a discrete cosine transform, is applied to the output of M filters, X_m. These filters are triangular and cover the $156 - 6844$ Hz frequency range; they are spaced on the Mel-frequency scale. This scale is logarithmic above 1 kHz and linear below this frequency. These filters are applied to the log of the magnitude spectrum of the signal, which is estimated on a short-time basis. The equation describing MFCCs is:

$$MFCC_n = \sum_{m=1}^{M} X_m \; cos\left(\frac{\pi \, n}{M}(m - 0.5)\right), n = 1, 2, ..., N \tag{6.7}$$

where N is the number of the cepstral coefficients, M is the analysis order and $X_m, m = 1, 2, ..., M$, represents the log-energy output of the mth filter. 20 triangular bandpass filters were used.

6.3.2   Robust Auditory-Based Phonetic Features

The human auditory model used here consists of three parts that simulate the behavior of the ear [18]. The external and middle ear are modeled using a bandpass filter that can be adjusted to signal energy in order to simulate the various adaptive ossicle motions. The inner part of the model simulates the basilar membrane (BM) that acts substantially as a non-linear filter bank. Due to the variability of its stiffness, different places along the BM are sensitive to sounds with different spectral properties. Actually, the BM is stiff and thin at the base, but less rigid and more sensitive to low frequency signals at the apex. Each location along the BM has a specific frequency, at which it vibrates maximally for a given input sound. This behavior is simulated thanks to the Caelen model by a cascade filter bank [18]. The bigger the number of these filters the more accurate is the model. Usually it is recommended to consider 24 filters. This number depends on the sampling rate of the signals and on other parameters of the model such as the overlapping factor of the bands of the filters, or the quality factor of the resonant part of the filters. The final part of the model deals with the electro-mechanical transduction of hair-cells and afferent fibers and the encoding at the level of the synaptic endings. Although, based on the physiological function of each module, the Caelen's auditory model is used with the purpose of encoding speech signals by accommodating the ear properties in order to extract pertinent phonetic features.

6.3.2.1 Mid-External Ear

The external and middle ear are modeled using a bandpass filter. The recurrent formula of this filter is the following:

$$s'(k) = s(k) - s(k-1) + \alpha_1 s'(k-1) + \alpha_2 s'(k-2) \tag{6.8}$$

where $s(k)$ is the speech wave, $s'(k)$ is the filtered output, $k = 1,...K$ is the time index and K the number of samples in a given frame. The coefficients α_1 and α_2 depend on the sampling frequency F_s, the central frequency of the filter and its Q-factor. The values of 1500 Hz as central frequency and 1.5 as Q-factor are convenient.

6.3.2.2 Mathematical Model of the Basilar Membrane

After each speech frame is transformed by the mid-external filter, it is passed to the cochlear filter banks whose frequency responses simulate those given by the BM for an auditory stimulus in the outer ear [18]. The formula of the model is as follows:

$$y_i(k) = \beta_{1,i} y_i(k-1) - \beta_{2,i} y_i(k-2) + G_i[s'(k) - s'(k-2)] \tag{6.9}$$

and its transfer function can be written as:

$$H_i(z) = \frac{G_i\left[1 - z^{-2}\right]}{1 - \beta_{1,i} z^{-1} + \beta_{2,i} z^{-2}} \tag{6.10}$$

where $y_i(k)$ is the BM displacement which represents the vibration amplitude at position x_i and constitutes the BM response to a mid- external sound stimulus $s'(k)$. The parameters G_i, $\beta_{1,i}$ and $\beta_{2,i}$, respectively the gain and coefficients of filter (also called channel) i, are functions of the position x_i along the BM. N_c cochlear filters are used to realize the model; N_c is set to 24 in our experiments. These filters are characterized by the overlapping of their bands and a large bandwidth. We assume that the BM has a length of 35 millimeters which is approximately the case for humans. Thus, each channel represents the state of an approximately $\Delta x = 1.46$ mm of the BM, which is the smallest unit physically simulated. An augmentation of the number of channels will reduce this basic unit size, which will improve the precision of the model but will increase the number of parameters.

> **Initialize** $f_x = (F_s \Delta x)^2$, $H_0 = 0$, $r_{i,j} = 0$; $E_0 = 0$.
> **For** $i = 1$ to N_c **do**
> $\quad x_i = i\Delta x$; $v = e^{(-106.5x_i)}$; $F_i = 7100v - 100$; $C_i = \frac{(27v)^2}{f_x}$;
>
> $\quad Q_i = (-8300x_i + 176.3)x_i + 4$; $G_i = e^{(-80x_i)}$; $u = e^{-\frac{\pi F_i}{F_s Q_i}}$;
>
> $\quad \beta_{1,i} = 2u\cos(\frac{2\pi F_i}{F_s})$; $\beta_{2,i} = u^2$;
>
> $\quad E_i = \frac{1}{1+(2-E_{i-1})C_i}$; $A_i = E_i C_i$;
> **EndDo**
> **For** $k = 1$ to K **Do**
> $\quad$ **For** $i = 1$ to N_c **Do**
> $\quad\quad H_i = \left(G_i(s'(k) - s'(k-2)) + \beta_{1,i}r_{i,2} - \beta_{2,i}r_{i,1}\right)E_i + H_{i-1}A_i$
> $\quad$ **EndDo**
> $\quad$ **For** $i = N_c$ to 1 **Do**
> $\quad\quad r_{i,3} = A_i r_{i+1,3} + H_i$, and $y_i'(k) = r_{i,3}$
> $\quad$ **EndDo**
> $\quad$ **For** $i = 1$ to N_c **Do**
> $\quad\quad$ **For** $j = 1$ to 2 **Do**
> $\quad\quad\quad r_{i,j} = r_{i,j+1}$
> $\quad\quad$ **EndDo**
> $\quad$ **EndDo**
> **EndDo**

Fig. 6.1 Sample-by-sample algorithm for extracting the cochlear signal with the hair cell and fiber effects. The parameter F_s is the sampling frequency fixed at 16000 Hz, N_c is the number of channels fixed at 24, F_i is the central frequency of each channel, G_i is the gain of filter i, C_i is a coupling coefficient, E_i is the direct coupling function, A_i is the inverse coupling function and H_i, $r_{i,j}$ are temporary calculation functions.

6.3.2.3   Hair Cells and Afferent Fibers

In order to not over-emphasize the problem of electro-mechanical transduction in hair cells and fibers, only the coupling effects they induce are taken into account in the Caelen ear model. Thus, the main feature of the model retained for hair cells and fibers is supplied by the C_i, E_i and A_i coupling parameters and used by the sample-by-sample algorithm described in Figure 6.1. $y_i'(k)$ provided by the algorithm can be regarded as the resulting stimulus after the passage through the mid-external ear, the basilar membrane with the effect of hair cells, and afferent fibers. This new set of samples is used to compute the amount of energy for each channel.

6.3.2.4   Encoding and Cues Extraction

The energy of the stimulus propagated through the nerve fibers along each portion Δx of the cochlea is calculated and lightly smoothed in order to be exploited for extracting pertinent information. The absolute energy of each channel is given by:

$$W_i'(T) = 20 \log \sum_{k=1}^{K} |y_i'(k)| \tag{6.11}$$

In Equation 6.11, T refers to the frame index. Between the current and previous frames, a smoothing function is applied to smooth the energy fluctuations. The smoothing equation is:

$$W_i(T) = c_0 W_i(T-1) + c_1 W_i'(T) \tag{6.12}$$

where $W_i(T)$ is the smoothed energy, and c_0 and c_1 are coefficients for averaging the terms $W_i(T-1)$ and $W_i'(T)$ such that the sum of the two coefficients is unity.

To achieve the encoding processing, acoustic distinctive cues are calculated starting from the data using linear combinations of the energies taken in the channels simulating the BM, hair cells and fibers. We have considered the classic distinctive feature set laid out by Chomsky and Halle [23] and earlier by Jakobson [71]. The established criterion to retain a particular feature is its discriminative power and the fact that all features must fully distinguish all sounds in the language. It was shown in [71] that 12 acoustic cues are sufficient to characterize acoustically all languages. However, it is not necessary to use all of these cues to characterize a specific language. In our study, we choose, in addition to the mid- external energy of the ear, 7 cues to be used as robust features in an attempt to improve the performance of ASR. These 7 cues are based on the Caelen ear model described above, which does not correspond exactly to Jakobson's cues. These seven normalized acoustic cues are: Grave/Acute (G/A), Open/Close (O/C), Diffuse/Compact (D/C), Flat/Sharp (F/S), Mellow/Strident (M/S), Continuant/Discountinuant (C/D) and Tense/Lax (T/L). They have been defined for each frame, as the following:

- **Grave/Acute:** is measured by taking the difference of energy between low frequencies within the (50-400 Hz) band and high frequencies within the (3800-6000 Hz) band, which corresponds to the following linear combination of particular channel energies:

$$G/A = (W_1 + ... + W_5) - (W_{20} + ... + W_{24}) \tag{6.13}$$

- **Open/Closed:** a given phoneme is considered closed if the energy of low frequencies (230-350 Hz) is greater than that of the middle frequencies (600-800 Hz). Hence, the O/C cue is calculated by:

$$O/C = W_8 + W_9 - W_3 - W_4, \tag{6.14}$$

- **Diffuse/Compact:** compactness is characterized by the prominence of the central formant region (800-1050 Hz) compared with the surrounding regions (300-700 Hz) and (1450-2550 Hz). The calculation of the D/C cue is as follows:

$$D/C = W_{10} + W_{11} - \left(W_4 + ... + W_8 + W_{13} + ... + W_{17}\right)/5 \qquad (6.15)$$

- **Flat/Sharp:** when the energy contained within (2200-3300 Hz) is more important than the one contained within (1900-2900 Hz), the event is considered as sharp. The F/S feature is given by:

$$F/S = W_{17} + W_{18} + W_{19} - W_{11} - W_{12} - W_{13} \qquad (6.16)$$

- **Mellow/Strident:** the principal characteristic of strident phonemes is the presence of noise due to a turbulence at their articulation point. Hence, a phoneme is considered as strident if the frequency range (3800-5300 Hz) contains more energy than the (1900-2900 Hz) frequency range, which is quantified by:

$$S/M = W_{21} + W_{22} + W_{23} - W_{16} - W_{17} - W_{18} \qquad (6.17)$$

- **Continuant/Discountinuant:** this cue quantifies the variation of the spectrum magnitude by comparing the energy of current and preceding frames. It will be low for slow variations and high when important fluctuations are encountered. It is calculated by:

$$C/D = \sum_{i=1}^{N_c} \left| W_i(T) - Wa(T) - W_i(T-1) + Wa(T-1) \right| \qquad (6.18)$$

where $W_i(T)$ and $Wa(T)$ are, respectively, energy of channel i and energy average over all channels of current frame T. The offset is removed by subtracting the energy average.

- **Tense/Lax:** is measured by taking the difference of energy between middle frequencies within the (900-2000 Hz) range and relative high frequencies within the (2650-5000 Hz) range, i.e.,

$$T/L = \left(W_{11} + ... + W_{16}\right) + \left(W_{18} + ... + W_{23}\right) \qquad (6.19)$$

Figure 6.2 gives an example of the evolution of the cues derived from Caelen's auditory model for the phrase: *She had your dark suit in greasy wash water all year*, taken from the TIMIT database.

The Caelen's distinctive cues (CDCs) are calculated starting from the spectral data using linear combinations of the energies taken in various channels. Indeed through such calculations one seeks to describe with these few parameters the spectral distribution and its temporal evolution.

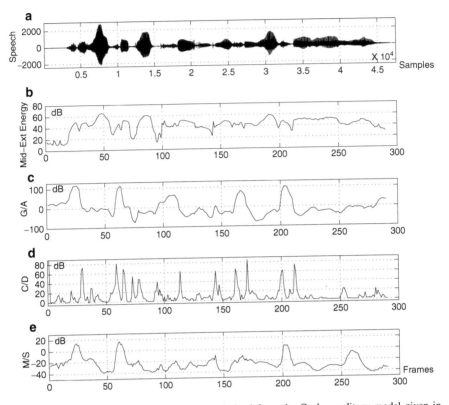

Fig. 6.2 Examples of distinctive feature cues derived from the Caelen auditory model given in dB. These cues are (b) Mid-external ear energy, (c) G/A, (d) C/D and (e) S/M. The sentence (a) is: 'She had your dark suit in greasy wash water all year', uttered by a female speaker.

6.4   ASR Robustness

Adaptation to the environment changes and artifacts remains one of the most challenging problems for speech recognition. As speech and language technologies are being transferred to real applications, the need for greater robustness in recognition technology becomes more apparent when speech is transmitted over telephone lines, when the signal-to-noise ratio (SNR) is extremely low, and more generally, when unpredictable acoustic conditions occur. A speech recognition system is considered robust if it maintains satisfactory recognition performance in adverse conditions. To cope with these adverse conditions and to achieve noise robustness, different approaches have been studied. Three major approaches have emerged: the signal compensation, feature space, and model adaptation techniques.

6.4.1 Signal compensation techniques

These methods rely on pre-processing the corrupted speech input signal prior to the pattern matching in an attempt to enhance the signal-to-noise ratio. The idea is to modify the noisy signal y with a transformation E(.) such that the distributions of the modified noisy signal $E(y) = z$ resemble those of x, the clean data used to train the recognizer. The goal is to transform the noisy signal to resemble clean speech as performed by the signal enhancement methods depicted in Part I. These types of techniques attempt to "clean" the distorted signal or apply an inverse transform, which reverses the effect of the distortion. Many different transformation operators have been proposed over the years. These operators may have different assumptions about the type of noise or other sources of degradation, and the type of features to be extracted from the signal for the recognition task. It is important to mention that the performance of some speech enhancement algorithms is more often evaluated in terms of recognition accuracy.

6.4.2 Feature Space Techniques

The performance of a robust speech recognition system is influenced by the ability of the acoustic features to represent the relevant information of spoken speech. Most of these acoustic features are known to be sensitive to noise and distortions and degrade the performance of a speech recognizer when deployed in adverse conditions. On the contrary, humans are capable of recognizing a speech uttered in very noisy conditions and affected by various channel distortions. It is therefore argued for the utilization of the human auditory system properties in the acoustical analysis. The most common auditory-based features used in current speech recognition systems are the cepstral front-end derived from a set of Mel-spaced filter banks presented in Section 6.3.1 and the Perceptual Linear Predictive analysis which utilizes several human auditory properties including the Bark frequency scale. Auditory-based methods that attempt to model the psychoacoustics and neurophysiology mechanisms show evidence that auditory properties are useful in robust speech recognition [71]. As we will show in Chapter 8, the ear model presented in Section 6.3.2 has also proven effective as an ASR front-end. Feature space techniques that aim at giving more robustness to speech recognition are concerned with the filtering or transformation of feature vectors. The goal of this filtering is to remove unwanted distortions so that the resulting features are "cleaner", i.e., their mismatch to clean speech models is reduced. This is performed by transforming the features $F(y)$ computed from the noisy speech that are modified by a transformation C(.) in such a way that the distributions of the transformed features $F_Z = C(F(y))$ better match the "canonical" distributions used by the recognizer. Recognition is then performed with the transformed features F_Z. There are several Feature space techniques that improve the ASR robustness, such as the

Cepstral Mean Subtraction (CMS), Cepstral Variance Normalization (CVN), Vocal Tract Length Normalization (VTLN) and histogram normalization and rotation [75, 13, 34].

6.4.3 Model Space Techniques

These techniques attempt to establish a compensation process which modifies the pattern matching itself to account for the effects of noise. Model compensation (or adaptation) methods assume that in noisy environments the clean (or reference) speech models are transformed by a $T(.)$ function. Compensation methods estimate for the inverse transform function $T^{-1}()$, and provide an estimation of the clean models. The transformation may be such that it modifies the basic structure of the statistical model (HMM), e.g., model decomposition, or merely modifies its state distributions without affecting the topology. Transformations that do not affect the model topology can be broadly divided into two categories: methods based on analytical characterizations of the effect of noise, such as Parallel Model Combination (PMC) [47] and Jacobian Adaptation (JA) [119] and methods based on empirical evidence obtained from noisy data. The latter type of methods can be categorized as those that modify parameters based exclusively on the empirical evidence obtained from noisy data, e.g., MLLR [84], and those that use *a priori* information about the statistical distribution of parameters of state distributions, such as MAP adaptation [83].

In the basic PMC process, a combination of a Gaussian clean speech model and noise model are combined in order to model the effect of additive noise. These models are expressed in the linear-spectral (referred to as *lin*) or log-spectral domains, assuming that the sum of two log-normally distributed variables is also log-normally distributed. The model parameters, namely the mean vectors and covariance matrices, are combined by using the following equations:

$$\hat{\mu}^{lin} = \gamma log(exp(\mu^{lin})) + log(exp(\tilde{\mu}^{lin})),$$
$$\hat{\Sigma}^{lin} = \gamma^2 log(exp(\Sigma^{lin})) + log(exp(\tilde{\Sigma}^{lin})) \tag{6.20}$$

where $(\hat{\mu}, \hat{\Sigma})$ and (μ, Σ) are the noisy and clean speech model parameters respectively, $(\tilde{\mu}, \tilde{\Sigma})$ are the noise model parameters, and γ is a gain matching term which determines the signal-to-noise ratio. The noisy speech is modeled with $N \times M$ states, where N states are used for clean speech, and M states for the noise. The Viterbi algorithm is used to perform simultaneous recognition of speech and noise. In the case of non-stationary noises, several states M can be used to model the noise. In the case of stationary noises, one state would be enough to represent the noise. The main drawback of this method is the computational load.

The Jacobian approach adapts the HMM models trained in the reference condition to the target noise condition. The noise change is expressed in terms of a Jacobian matrix. To achieve effective recognition performance the target and reference conditions should be close and the change in the statistics of noisy speech should be in the linear range of the Jacobian approximation. The feature of JA is that the non-linear transformation of models caused by the fluctuation of environment is approximated as linear in cepstrum domain:

$$Y_{tar}^c \cong Y_{ref}^c + J_N(N_{tar}^c - N_{ref}^c), \tag{6.21}$$

where Y_{ref}^c and N_{ref}^c are speech and noise of reference environment, and Y_{tar}^c and N_{tar}^c are speech and noise of target environment, respectively. Y^c represents vectors of cepstrum domain. The Jacobian matrix J_N can be calculated as follows:

$$J_N = \frac{\partial Y^c}{\partial N^c} \tag{6.22}$$

$$= C\frac{N^s}{Y^s}C^{-1}, \tag{6.23}$$

where C is the cosine transform matrix. Y^s and N^s represent vectors in the spectra domain. It should be noted that the former equations deal with the feature space adaptation. This framework is also used to adapt the HMM's stochastic parameters under the assumption that the variance of the Y^c distribution is sufficiently small. The mean vector and covariance matrix of each distribution of HMMs can be adapted to the new environment; only the adaptation of the mean vector is considered in practical implementations. This is because the adaptation of covariance matrices did not show significant improvement in the recognition performance.

$$\hat{\mu}_{tar} = \hat{\mu}_{ref} + J_N(\tilde{\mu}_{ref} - \tilde{\mu}_{tar}). \tag{6.24}$$

HMMs are firstly trained with data of reference environment. In most JA methods, the PMC algorithm is used in the training phase, then the adaptation of the models is performed by using Equation 6.24. JA is considered as a fast model adaptation technique for a new acoustical environment and numerous variants have been proposed in the literature [74].

Maximum likelihood linear regression (MLLR) and Maximum a Posteriori (MAP) are also popular methods for adapting to a new environment. These methods were originally developed for speaker adaptation, but they have also been used to carry out environmental compensation. The widely-used adaptation technique is MLLR [84] [95]. It is a parameter transformation technique that has proven successful while using a small amount of adaptation data. It computes a set of transformations that will reduce the mismatch between an initial model set and the adaptation data. MLLR is a model adaptation technique that estimates a set of linear transformations for the mean of Gaussian mixture HMM system. The effect of these transformations is to shift the component means in the initial system so that each state in the HMM is more likely to generate the adaptation data. The principle of mean transform in the MLLR scheme, assumes that Gaussian mean vectors are updated by linear transformation. Let μ_k be the baseline mean vector and $\hat{\mu}_k$ the corresponding adapted mean vector for an HMM state k. The relation between these two vectors is given by: $\hat{\mu}_k = \mathbf{A}_k \xi_k$ where $\mathbf{A}_k$ is the $d \times (d+1)$ transformation matrix and $,_k = [1, \mu_{k1}, \mu_{k2}, ..., \mu_{kd}]^t$ is the extended mean vector. It has been shown in [84] that maximizing the likelihood of an observation sequence o_t is equivalent to minimizing an auxiliary function Q given as follows:

$$Q = \sum_{t=1}^{T} \sum_{k=1}^{K} \gamma_k(t)(o_t - \mathbf{A}_k \xi_k)^T C_k^{-1} (o_t - \mathbf{A}_k \xi_k), \tag{6.25}$$

where $\gamma_k(t)$ is the probability of being in the state k at time t, given the observation sequence o_t. C_k is the covariance matrix which is supposed to be diagonal. The general form for computing optimal elements of $\mathbf{A}_k$ is obtained by differentiating Q with respect to $\mathbf{A}_k$:

$$\sum_{t=1}^{T} \gamma_k(t) C_k^{-1} o_t \xi_k^t = \sum_{t=1}^{T} \gamma_k(t) C_k^{-1} \mathbf{A}_k \xi_k \xi_k^t. \tag{6.26}$$

Depending on the amount of available adaptive data, a set of Gaussians, and more generally, a number of states will share a transform, and will be referred to as regression class r. Then, for a particular transform case $\mathbf{A}_k$, Gaussian components will be tied together according to a regression class tree and the general form of 6.26 expands to:

$$\sum_{r=1}^{R} \sum_{t=1}^{T} \gamma_{k_r}(t) C_{k_r}^{-1} o_t \xi_{k_r}^t = \sum_{r=1}^{R} \sum_{t=1}^{T} \gamma_{k_r}(t) C_{k_r}^{-1} \mathbf{A}_k \xi_{k_r} \xi_{k_r}^t. \tag{6.27}$$

In standard MLLR, the column by column estimation of $\mathbf{A}_k$ elements is given as follows:

$$a_i = G_i^{-1} z_i, \tag{6.28}$$

where z_i refers to the ith column of the matrix which is produced by the left hand side of 6.27, and where G_i is given by $\sum_{r=1}^{R} \hat{c}_{ii}^{(r)} \xi_{k_r} \xi_{k_r}^t$, where $c_{ii}^{(r)}$ is the ith diagonal element of $\sum_{t=1}^{T} \gamma_{k_r}(t) C_{k_r}^{-1}$.

In the application described in Chapter 9, only one regression class is used.

The system adaptation can also be accomplished using the Maximum *a posteriori* (MAP) technique [83]. For MAP adaptation, the re-estimation formula for Gaussian mean is a weighted sum of the prior mean with the maximum likelihood mean estimate. It is formulated as:

$$\hat{\mu}_{ik} = \frac{\tau \mu_{ik} + \sum_{t=1}^{T} \varphi_t(i,k) x_t}{\tau + \sum_{t=1}^{T} \varphi_t(i,k)}, \qquad (6.29)$$

where τ_{ik} is the weighting parameter for the kth Gaussian component in the state i, and $\varphi_t(i,k)$ is the occupation likelihood of the observed adaptation data x_t. One of the drawbacks of MAP adaptation is that it requires more adaptation data to be effective compared to MLLR. Generally speaking, model-based approaches often perform better than feature-based approaches, normally at the cost of higher computational complexity.

6.5   Speech Recognition and Human-Computer Dialog

The primary goal of speech recognition is to provide an alternative to text input; therefore, subsequent semantic or pragmatic analysis is usually considered as language or speech understanding. The role of the spoken language understanding (SLU) system is to infer users' intentions from speech, and do it robustly when spontaneous speech effects (hesitation, self correction, stuttering,...) occur. One increasingly popular approach to cope with these issues consists of extending the statistical pattern recognition framework, commonly used for speech recognition, to the SLU problem. For this purpose, a pattern recognition based SLU relies on the semantic language model (SLM) to detect semantic objects and construct a parse tree from users' utterances. This SLM is usually realized so that the semantic structure of the utterance can be included in a dialog manager component. Current speech-enabled interfaces tend to give an increasing role to the dialog manager in order to improve the interaction naturalness. In these configurations, the role of the speech recognizer is to provide n-best lists or lattices to the natural language parser (language model) which outputs the most plausible interpretation and then passes it to the dialog manager for making a final decision according to the current context. In this section, we will discuss some new strategies to overcome speech recognizer errors in order to enhance the performance and quality of the dialog experience.

Speech-enabled interfaces fall into three broad categories. The first category includes Command and Control (C&C) interfaces that rely on a fixed task dependent grammar to provide user interaction [100]. Their main advantage is their ease of implementation and high command recognition rate. However, their downside is the high cognitive load required to learn and use the system because of its lack of flexibility. The second category is based on interactive voice response (IVR) that guides users by using prompts in order to validate the utterance at every step [126]. This style of interaction is mostly used in menu navigation such as that found with phone and cable companies. Its relative lack of efficiency for fast interaction makes it a poor choice for every day use. Finally, the third category uses natural language (NL) processing to parse the user's utterance and to determine the goal of the request. This can be done through multiple ways such as semantic and language processing (SLM). Systems belonging to this category are characterized by the complexity of parsing spontaneous utterances that might not follow conventional grammars. In this section, we will present a practical application involving a dialog system belonging to this latter category.

6.5.1 Dialog Management Systems

Efficiency and effectiveness of dialog systems remain below expectations because they assume that recognizers do not make errors. Therefore, the main challenge facing SLU systems is to overcome the errors from ASR [46]. Various dialog frameworks have been experimented to limit the impact of the recognizers' errors. To face the challenge, recent approaches mostly focus on using the dialog knowledge itself.

Current dialog models are governed by inference rules and use discourse constraints, domain knowledge, and applicable conditions to build a 'belief' state that helps the system providing the best answer. For instance, Gorrell in [52] categorized the answers of the SLU into twelve categories based on the patterns of speech recognition errors, and selected the most likely answer by using SLM. Kang et al. propose dialog strategies that allow the system to share knowledge with a user during a dialog [76].

One solution proposed by Visweswariah and Printzis [139] is based on the dialog context or state-dependent language modeling that performs interpolation of the n-gram language model and dialog specific models. A feedback to the parser is maintained during the dialog thanks to the dialog "state". This state represents the situation in which the dialog is. This dialog state information considers using the prompt that the system uses to constrain the language model. A complete synergy is then built between the dialog system and the language model to overcome the misunderstandings.

Another solution proposed by Hacioglu and Ward [55] consists of splitting language modeling into dialog-dependent modeling using n-grams and syntactic modeling based on a set of stochastic context-free grammars. This approach involves four integrated models: the concept model, the syntactic model, the pronunciation model and the acoustic model. The concept model represents the a priori probabilities of concept sequences conditioned on the dialog context. The syntactic model is the probability of word strings used to express a given concept. The pronunciation model and the acoustic model give the probabilities of possible phonetic realizations and the acoustic feature observations respectively. The reported results show a significant perplexity improvement.

The most investigated framework is based on Partially Observable Markovian Decision Processes (POMDPs). The Markovian Decision Processes (MDPs) were applied for the first time to dialog management by Levin and Pieraccini [85] and Singh et al. [128], who have implemented them in a real system. The main limitation of MDPs is that they assume that the conversation is known exactly and thus, they do not consider the probable inefficiency of the recognition systems in some

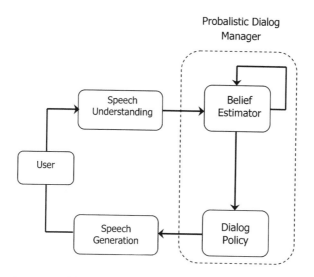

Fig. 6.3 General structure of the POMDP-based dialog system.

situations. This limitation leads many researchers to the development of dialog systems based on the POMDPs that have the advantage of expressing uncertainty in the current state of the 'conversation' [101,117,59,146,151]. In these configurations, the different sources of degradations, such as the speech recognition errors, are considered. They also benefit from observations provided by a variety of sources such as acoustic confidence and parsing score. This approach requires that a belief state be maintained during the conversation. As illustrated in Figure 6.3, from the paper of Young *et al.* [151], the principle of POMDP-based dialog consists of capturing the users last input dialog act, the users goal, and a record of the dialog history. The belief estimator manages the uncertainty by providing values of the belief state that permit the dialog policy module to determine the next action to be performed by the system. The dialog policy can be optimized by assigning weights to actions and states. This gives the dialog manager the capacity to deal with uncertainty and to accommodate the *n*-best recognizer outputs. The POMPDs continue to focus researchers' interest and the framework is still under improvement as shown in [48, 151].

6.5.2 Dynamic Pattern Matching Dialog Application

To perform a natural human-system spoken interaction in a realistic application, we proposed in [11] a solution which consists of providing the speech modality to allow human operators of a Radio Frequency IDentification (RFID) Network to communicate naturally with related devices and information systems. The

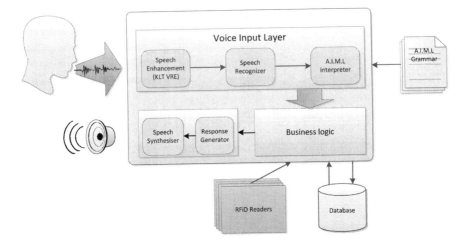

Fig. 6.4 Spoken dialog system interacting with RFID Network.

advantages of speech hands-free and eyes-free systems combined with the RFID technology are expected to improve speed, ergonomics and safety of numerous operations such as order picking, quality management, tracking and monitoring of assets, and shipping operations.

As shown in Figure 6.4, a spoken dialog system is integrated in the RFID-based application in order to help a human operator to perform multiple and simultaneous tasks including the use of a number of devices and to update information systems on the ground. The workflow may include other additional tasks such as inspecting items received for damages, confirming the quantity of a given product, etc. To reduce the cognition load of the operator and to improve his/her comfort, the sensor capabilities of RFID are exploited. The application involves a robust ASR system using the KLT-VRE speech enhancement method presented in Chapter 4. The n-best ASR recognition outputs are then fed to the dialog manager which parses them and returns either a user-friendly message to the user to indicate that it did not understand the meaning of the utterance or the most likely command with its parameters to the business logic end. The proposed system enables the orchestration of RFID events and inputs into synchronized operations in order to provide more automation thanks to the spoken dialog interpreter.

The dialog interpreter is based on an Artificial Intelligence Markup Language (AIML) parser [4]. AIML is an XML compliant language designed to create chat bots. Its main characteristic is minimalism since it can reduce complex user utterances into simpler atomic components. This framework can be considered as a pattern matching system that maps well with Case-Based Reasoning. AIML consists of categories that encapsulate a pattern, a user input, a template, and the possible answer. The parser then tries to match what the user said to the most likely pattern and provides the corresponding answer. This system supports recursion, which enables the dialog manager to provide answers based on previous inputs. Additionally, patterns can include wildcards that are especially useful to infer the user's goals and to update the belief state. Additional details about this AIML-based spoken dialog system can be found in [125].

6.6   ASR Robustness and Soft Computing Paradigm

ASR Robustness is essential to maintain a high level of performance regarding the wide variety of dynamically changing acoustic environments in which a speech recognition system must inevitably operate. The acoustic model, the language model and eventually the spoken language understanding module are used to capture the speech information at different levels. To face the mismatch between the training and testing environments many methods that target the robustness of speech recognition have been developed. In the field of ASR robustness, the systems have to deal with uncertainty. It is clear from experience that the systems cannot run on precise decisions that are the results of the versatility of speech, behavior of speakers, and environment. It should be noted that almost all proposed approaches make restrictive (sometimes unrealistic) assumptions in order to overcome the complexity of the problem. Approximations, constraints and limitations are often used to model the robustness problem. The proposed models proceed through this situation by accepting the obtained solutions regardless of the restrictions they made. To deal with this uncertainty and imprecision, soft computing appears to be a promising approach which could be used complementarily with the signal compensation, feature space, and model adaptation techniques. According to the definition given by Professor L. Zadeh "The guiding principle of soft computing is to exploit the tolerance for imprecision, uncertainty, partial truth, and approximation to achieve tractability, robustness and low solution cost and better rapport with reality" [127].

Our vision for the integration of soft computing techniques in the ASR robustness paradigm is illustrated in Figure 6.5. The acoustic mismatches may occur in signal, feature and model spaces between the training and the testing environments. Let X denote the space of the raw speech signal in the training environment. The mismatch between the training and testing environments is modeled by a distortion $E(.)$ which transforms X to Y. In speech recognition, feature extraction is carried out. These features are represented as F_X and F_Z in the training and testing environments, respectively. The mismatch between the two environments in the feature space is modeled by the function $F(.)$, which transforms the features from F_X to F_Z. Finally, the features are used to build models, HMMs in our case. The mismatch between the training and testing environments can be viewed in the model space as the transformation $T(.)$ that maps Λ_X to Λ_T. Sources of mismatch may include additive noise, channel and transducer differences, and speaker variability such as regional accents, speaking rates and styles. Soft computing techniques have proven very effective in optimization problems. We investigate their ability to generate a

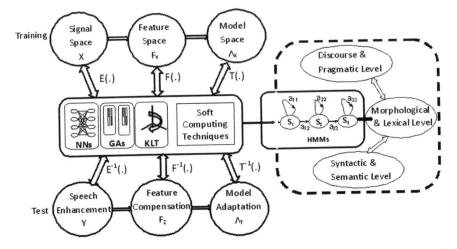

Fig. 6.5 A new vision towards the implication of soft-computing techniques in robust and natural speech recognition and understanding systems.

diverse set of solutions for robust features or models using soft conditions. The natural processes that are the inspiration behind the design of neural networks and evolutionary algorithms coupled with subspace decomposition frameworks are exploited to optimize the transformation functions: $E(.)$, $F(.)$ and $T(.)$. In the subsequent chapters, we will discuss this new paradigm and related techniques.

6.7 Summary

One of the hottest topics of Human-Computer Interaction research targets the realization of a natural, intuitive and multimodal interaction where the speech modality plays a central role. However, current interaction frameworks using speech recognition are still difficult to use. They still require a quiet environment and a long training phase to reach fully effective and optimal use. Despite the prodigious advances of ASR systems, recognition errors will occur and cannot be handled only by the language model. Indeed, it is not possible for speech recognition to cope with all possible sources of error, such as strong accents, noisy channels, background noise, spontaneous utterances and also the use of words that do not occur in the system's vocabulary. In order to counter these limitations, soft computing techniques present an interesting solution. The next chapters will present some practical solutions showing the advantages of using soft computing to improve ASR robustness.

Chapter 7
Artificial Neural Networks and Speech Recognition

Abstract The ability to solve some classification problems with relative ease and without the need to formalize the statistical properties of the problem have made neural networks very popular in the field of speech processing. In this chapter, the usefulness of a neural network using autoregressive backpropagation and time-delay components (AR-TDNN) is illustrated. Combined with HMMs, the AR-TDNN is incorporated in a flexible hybrid structure which attempts to improve the performance of complex phonetic feature (nasality) detection and classification.

Keywords Hybrid speech recognition • Hidden Markov Models • Time-Delay neural networks • Hierarchical connectionist structure • Autoregressive backpropagation • French nasal vowels

7.1 Related Work

Speech recognition is basically a pattern recognition problem. Since neural networks have demonstrated their success in pattern recognition, numerous earlier studies naturally applied neural networks to speech recognition. These studies involved simplified tasks such as voiced/unvoiced or vowel/consonant classifications. The success in performing these tasks has motivated researchers to move to phoneme and isolated word identifications. The leading work of Alex Waibel demonstrates that neural networks, when they integrate the temporal component, can form complex decision surfaces from speech data [140].

Hybrid approaches that combine the discriminative capabilities of neural networks and the superior abilities of HMMs in time alignment were developed in the mid-nineties and continue to attract the interest of researchers [56, 148]. In the hybrid approach proposed by Rogoll in [115], the component modeling the emission probabilities of the HMM is replaced by a neural net. The most effective hybrid system was proposed by Bourlard and Morgan [15] and has shown that neural networks can be trained so that the output of the $i-th$ neuron estimates the posterior

S.-A. Selouani, *Speech Processing and Soft Computing*, SpringerBriefs in Electrical and Computer Engineering, DOI 10.1007/978-1-4419-9685-5_7,
© Springer Science+Business Media, LLC 2011

probability $p(W_i|x)$. In fact, the well-known hybrid speech recognition systems try to improve the discrimination performance of conventional HMMs while still adhering to the general statistical formalism. Neural network classifiers are naturally discriminative and do not impose constraints such as uncorrelated feature coefficients which make them useful in many complementary tasks besides the statistical-based systems. Despite their efficiency, current hybrid speech recognition systems require a great amount of time to train the neural networks. Also, they need to determine parameters such as the learning rate, momentum or batch size. Sizes of speech corpora used in various applications increase every day and therefore the training times for such systems become prohibitive, which practically limits their use.

7.2 Hybrid HMM/ANN Systems

Architectures of current ASR systems, mainly based on HMM of Gaussian mixtures, are compact and tackle the global recognition task head on. Common assumptions like state conditional observation independence and time independent transition probabilities, limit the classification abilities of HMMs. The monolithic approach they adopt, limits in some cases, the recognition performance, particularly when they are faced with complex features and/or prosody variations [50]. To face this drawback, this chapter investigates an alternative approach that consists of favouring a hybrid modular architecture of ASR rather than a monolithic one by using a mixture of neural network experts. The basic principle behind this modular structure is the well-known technique which consists of solving a complex problem by dividing it into simpler problems for which solutions can easily be obtained. These partial solutions are then integrated to provide an overall solution. In this context, we can cite the system described in [91] which is composed of two parts: the first part consists of an HMM involved in the recognition of specific phoneme classes, and the second part is composed of neural networks trained for the disambiguation of pairs such as the /m, n/ nasals. The results showed that significant improvements of ASR scores were obtained for both English and French. In their article, Hagen and Morris reviewed several successful extensions to the HMM/ANN in noise robust automatic speech recognition [56]. They presented various combinations of multiple connectionist experts, where each expert has different error characteristics that improve robustness to unpredictable signal distortion according to three main schemes. These schemes provide three different ANN and HMM combinations:

- *Feature combination* aiming at concatenating the data features from various sources and then using them as inputs to the ANN prior to HMMs.
- *Posterior probabilities combination* posterior probabilities estimated from each ANN are combined into a single set of probabilities.

- *Hypothesis combination* multiple sequence hypotheses are generated from systems with different error characteristics and are combined by using fusion methods such as recognizer output voting error reduction (ROVER) technique [44].

The solution presented in this chapter consists of using a hierarchical structure of neural experts as post-processors of HMM-based systems. This configuration seems well suited to exploit the discriminating capacities of neural networks. The final configuration is intended to be more flexible in order to be able to easily generalize the identification of complex features. To give additional discriminability for speech pattern comparison, an inclusion of hearing/perception knowledge is carried out through the use of auditory-based cues.

7.3   Autoregressive Time-Delay Neural Networks

Because speech is a highly dynamic and variable phenomenon, we consider Recurrent Networks (RNs) to be more adapted than feedforward networks in the case of any classification task dealing with speech. RNs are generally trickier to work with, but they are theoretically more powerful when dealing with dynamic events. They have the ability to represent temporal sequences of unbounded length. A good example of a recurrent net which takes into account the phonetic context effects is proposed by Russel [118] and uses an Autoregressive (AR) version of the backpropagation algorithm. This type of network is theoretically capable of capturing the coarticulation phenomenon of speech. However, even if RNs using AR perform very well in the context-dependent identification, their time alignment capability remains insufficient to tackle the phoneme length variability. To face this drawback, we can consider the integration, in conjunction with the AR component, of a delay component similar to the one proposed by Waibel through the Time-Delay Neural Networks (TDNN) [140]. This combination (AR-TDNN) is expected to be more powerful to discern some complex phonetic features even in a strong coarticulation context.

The autoregressive model proposed by Russel [118] includes an autoregressive memory which constitutes a form of self-feedback where the output depends on the current output and a weighted sum of previous outputs. Thus, the classical AR node equation is given by:

$$y_i(t) = f\left(bias + \sum_{j=1}^{P} w_{i,j} x_j(t)\right) + \sum_{n=1}^{M} a_{i,n} y_i(t-n), \qquad (7.1)$$

where $y_i(t)$ is the output of node i at time t, $f(x)$ is the $tanh(x)$ bipolar activation function, P is the number of input units, and M is the order of autoregressive prediction. Weights $w_{i,j}$, biases, and AR coefficients $a_{i,n}$ are adaptive and are optimized in order to minimize the output error. The AR-TDNN configuration

Fig. 7.1 AR-TDNN unit.

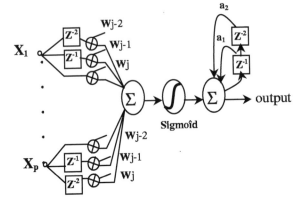

Fig. 7.1 AR-TDNN unit.

consists of incorporating a time delay component on the input nodes of each layer and then Equation 7.1 becomes:

$$y_i(t) = f\left(bias + \sum_{m=0}^{L} \sum_{j=1}^{P} w_{i,j,m} x_j(t-m)\right) +$$

$$+ \sum_{n=1}^{M} a_{i,n} y_i(t-n), \qquad (7.2)$$

where L is the delay order at the input. Feedforward and feedback weights were initialized from a uniform distribution in the range $[-0.9, 0.9]$. A neuron of the AR-TDNN configuration is shown in Figure 7.1. An autoregressive backpropagation learning algorithm performs the optimization of feedback coefficients in order to minimize the mean squared error $E(t)$ and defined as:

$$E(t) = \frac{1}{2} \sum_{i} \left(d_i(t) - y_i(t)\right)^2, \qquad (7.3)$$

where d_i is the desired value of the ith output node. The weight and feedback coefficient changes, noted respectively $w_{j,i,m}$ and $a_{i,n}$, are accumulated within an update interval $[T_0, T_1]$. In the proposed AR-TDNN version, the update interval $[T_0, T_1]$ is fixed such that it corresponds to the time delay of the inputs. The updated feedback coefficients are written as follows:

$$a_{i,n}^{new} = a_{i,n}^{old} + \frac{1}{T_1 - T_0} \sum_{t=T_0}^{T_1} \Delta a_{i,n}(t), \qquad (7.4)$$

and if T is the frame duration, the weights are as follows:

$$w_{i,j}^{new} = w_{i,j}^{old} + \frac{1}{LT} \sum_{t=T_0}^{T_1} \Delta w_{i,j}(t). \qquad (7.5)$$

The calculation of $\Delta a_{i,n}(t)$ variation is detailed in [118]. The optimization of weights and biases are performed as in Waibel's network [140]. Hence, the $\Delta w_{i,j}$ variations are accumulated during the update interval after accumulating the time-delay frames at the input.

7.4   AR-TDNN vs. TDNN

The Nguyen-Widrow initialization conditions are used for the initialization of the AR-TDNN [98]. The input layer receives the parameters from three frames. Each neuron of the hidden layer receives inputs from the coefficients of the three-frame window of the input layer. Cross-validation experiments using approximately 11852 phonemes uttered by four speakers (two males and two females) are carried out in order to compare the performances of AR-TDNNs and TDNNs. The task was to discriminate between emphatic, geminated vs. non-emphatic and non geminated consonants in the Arabic language [121]. The results given in Figure 7.2 show that the AR-TDNNs outperform the standard feedforward neural networks in all cases of complex phonetic feature classification. This experiment confirms the ability of the AR-TDNNs to perform both context-sensitive decisions and temporal component capture.

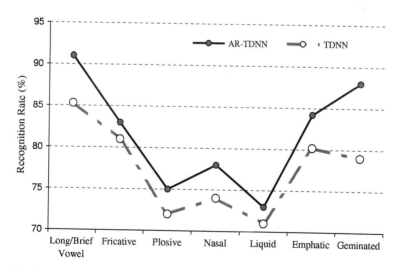

Fig. 7.2 Comparison of TDNN and AR-TDNN performances over macro-classes.

7.5 HMM/AR-TDNN Hybrid Structure

The training of the hybrid HMM/AR-TDNN system is carried out in two phases. The first phase involves the HMM performing an optimal alignment between the acoustic models of phones and the speech signal. The second phase consists of refining the HMM results by the AR-TDNN system. As illustrated in Figure 7.3, the global task is divided into two subtasks. For instance, the first subtask requires the HMM to achieve phone identification without discriminating between nasalized vs. oral vowel phones in French. The HMM will present a single label for the oral vowel and its nasalized counterpart. In the case of /a/ oral vowel and /a~/ nasalized vowel, a unique /A/ label is given. The /A/ sequence of phones is presented to the AR-TDNN system which makes final and finer decisions related to the nasalized/oral vowel discrimination.

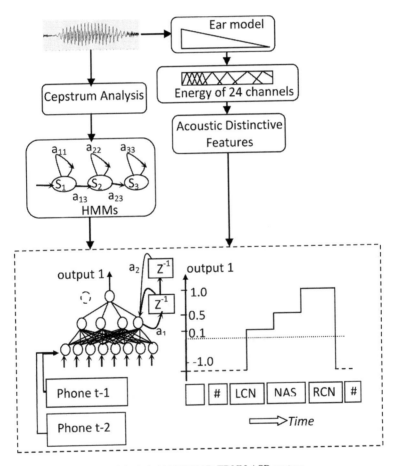

Fig. 7.3 General overview of the hybrid HMM/AR-TDNN ASR system.

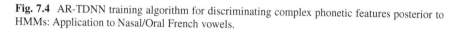

```
Initialize the output unit to (-1)
For i = 1 to Nphones do
    Scroll non discriminated phones
        If LCN is at input then output is set to (+0.1)
        EndIf
    Phone=Phone+1
    If NAS is at input then output is set to (+0.5)
        EndIf
    Phone=Phone+1
    If RCN is at input then output is set to (+1)
        EndIf
    Phone=Phone+1
EndDo
```

Fig. 7.4 AR-TDNN training algorithm for discriminating complex phonetic features posterior to HMMs: Application to Nasal/Oral French vowels.

The supervision of AR-TDNN learning considers phones as complete items that appear gradually in a given phonetic context. A specific training algorithm was developed so that, if a given phone context (the one we want to be learned) appears in the speech continuum, the AR-TDNN output activation increases gradually. In the example of the nasalization detection/classification, the task of AR-TDNN consists of learning to recognize the following sequence: LCN-NAS-RCN, where LCN is the left phonetic context of the nasalized vowel (referred to as NAS) and RCN is its right phonetic context. *NAS_NET* (Nasalization expert network) receives three input tokens at a time t and it should detect the nasalized sequence according to the algorithm given in Figure 7.4.

The learning sets the output at the high level (+1) when the end of the LCN-NAS-RCN sequence is attained. The low level (−1) is set otherwise, i.e. if a scrolling (stream) of oral vowel is observed. An autoregressive order of 2 is chosen and a delay of 2 phones is also fixed. These lower values of delay and order are justified by the fact that phones are used instead of frames. Other AR-TDNN-based expert systems can be provided. They can perform discrimination and identification of various phonetic features present in different languages. These tasks can be accomplished according to the same training protocol described in Figure 7.4.

7.6 Experiment and results

The purpose of this experiment is to investigate the detection and discrimination of vowel nasality in the French language. Nasal vowels are among the most recognizable features of a French accent. Nasality is a complex feature phenomenon which has been widely studied [21, 30, 109]. This feature results from the acoustic coupling the nasal cavities and the pharyngo-oral tract during speech. French has

Fig. 7.5 International
Phonetic Alphabet
representation of the four
French nasal vowels.

/ɑ̃/	[sɑ̃]	*sans*	'without'
/ɔ̃/	[sɔ̃]	*son*	'his'
/œ̃/	[bʁœ̃]	*brun*	'brown'
/ɛ̃/	[bʁɛ̃]	*brin*	'sprig'

four different nasal vowels. In Figure 7.5, the International Phonetic Alphabet (IPA) representations of the French nasal vowels are provided. Nasal vowels are produced with a lowering of the velum so that air escapes both through the nose and the mouth. The lowering of the velum affects the level of energy of a vowel and therefore nasal vowels have less energy when compared to oral vowels. It can also be observed that due to the nasalization, the first formant increases and the second formant decreases. The results of the imaging study carried out by Delvaux [31] show that the difference between nasal and oral vowels in French relies not only on the lowering of the velum, but also on many other characteristics such as lip rounding, tongue backing and tongue lowering.

7.6.1 Speech Material and Tools

The speech material was extracted from BDSONS, a French speech corpus consisting of 32 speakers: 16 male and 16 female. It is composed of bi-syllabic logatomes, numbers, digits, letters, and names (spelled in isolation and in connected speech) [19]. The acoustic subset (12 speakers) contains 600 CVCV including 20 consonants and semi-consonants and vowels; 200 consonant clusters; rhyme tests for consonant and vowels (pairs and triplets); 52 phonetically balanced sentences; 44 nasal sentences; 192 sentences including real words in French. The experiments are done on the syllable corpora of BDSONS. The test involves 840 nasal vowels with an equal number of each vowel (some additional recordings were necessary to reach this number of utterances).

The speech recognition by HMMs was performed by using the HTK toolkit described in [66]. HTK is an HMM-based speech recognition system. The toolkit can be used for either isolated or continuous whole-word/phone-based recognition. It supports continuous-density HMMs with the possibility to configure the number of states and mixture components. The MFCCs coefficients are used by HMMs as acoustical features. Twelve MFCCs are calculated on a 30-msec Hamming window advanced by 10 msec each frame. The log-energy and dynamic coefficients (e.g. first and second derivatives of MFCCs) are added to the static vector to constitute a 39-dimensional vector upon which the HMMs are trained. The baseline system for the recognition task uses 8-Gaussian mixture HMM system.

7.6.2 Setup of the Classification Task

NAS_NET, the nasalization expert based on AR-TDNN uses the cues derived from the ear model described in Section 6.3.2. In this model, the internal ear is represented by a coupled filter bank where each filter is centered on a specific frequency. Twenty-four filters (channels) are used. From the outputs of these channels, 7 cues are derived: acute/grave (AG), open/closed (OC), diffuse/compact (DC), sharp/flat (SF), mellow/strident (MS), continuous/discontinuous (CD) and tense/lax (TL). As shown in Section 6.3.2, these cues are very relevant to characterize the indicative features of many languages [71] over each identified phone generated by the HMMs thanks to the Viterbi algorithm (alignment procedure). The average of the ear-based indicative features is calculated over the frames composing the phone. The resulting average vector composed of 7 indicative features constitutes the AR-TDNN input vector. This vector is expanded by a component representing the middle ear energy extracted from the ear model. Thus, an 8-dimensional vector is used by the AR-TDNN expert. In Figure 7.6, three phones (sequences) are captured to illustrate the

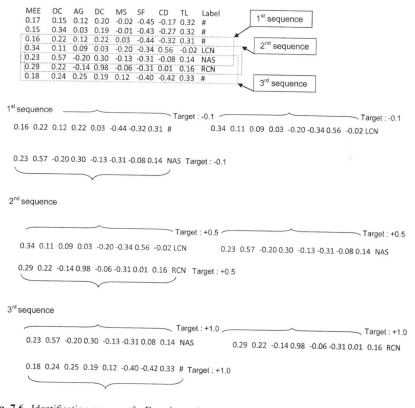

Fig. 7.6 Identification process of a French nasal vowel performed by the AR-TDNNs using 8 cues as inputs: middle ear energy (MEE), open/closed (OC), acute/grave (AG), diffuse/compact (DC), mellow/strident (MS), sharp/flat (SF), continuous/discontinuous (CD) and tense/lax (TL).

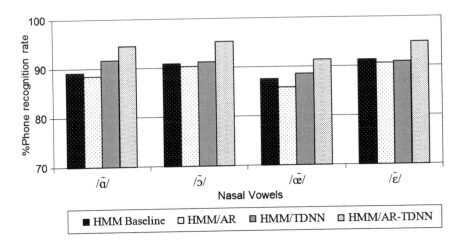

Fig. 7.7 General overview of the hybrid HMM/AR-TDNN ASR system.

process of the nasal vowel identification by AR-TDNNs. During the AR-TDNN training, the desired output (target) is set to −0.1 when the pattern of the first sequence is at input. When the second sequence passes through the input, the desired output is set to 0.5. The third sequence confirms that the nasal vowel, with its contexts, passed the network and therefore the desired output is set to +1.0. Otherwise the AR-TDNN output is set to −1.0. During the test phase, the AR-TDNNs are required to recognize the nasal vowel patterns.

7.6.3 Discussion

We compare the ability of the hybrid HMM/AR-TDNN system and baseline systems to perform the Nasal/Oral discrimination. Three baseline systems are considered. The first system is based only on HMMs and carries out the complete phonemic recognition of both oral and nasal vowels. The second system (HMM/AR) uses a recurrent neural network based on the autoregressive backpropagation algorithm as an expert for the discrimination between oral and nasal vowels. The third baseline system (HMM/TDNN) involves HMMs and TDNN to perform the oral/nasal disambiguation. The analysis of the results presented in Figure 7.7 reveals that the HMM/AR-TDNN configuration is the most accurate in the task of detection and identification of French nasal vowels. The HMM/AR-TDNN system achieves 94.1% average accuracy, while HMM, HMM/AR and HMM-TDNN achieve an average accuracy of 89.67%, 88.75%, and 90.5% respectively. We must underline the fact that the improvement reached by the structural modification of the ASR system is more significant than the inclusion of dynamic features such as first and second derivatives of MFCCs for instance.

7.7 Summary

In this chapter, a hybrid approach for the identification of complex phonetic features was presented. The objective was to test the ability of a system combining HMM and neural networks e.g. AR-TDNN to detect features as subtle as the nasalization feature of French vowels. For this particular task, it seems clear that the proposed hybrid HMM/AR-TDNN approach significantly improves the performance of standard HMMs. Dividing the global speech recognition task into subtasks assigned to complementary systems, conjugated with the use of dynamic ear-based distinctive features, constitutes a promising way to overcome issues related to language specificities. The case study presented in this chapter showed that using a soft computing technique as a post-process to the partially HMM recognized speech leads to a better performance than exclusively using the HMM processing.

Chapter 8
Evolutionary Algorithms and Speech Recognition

Abstract In this chapter, we present an approach for optimizing the front-end processing of ASR systems by using Genetic Algorithms (GAs). The front-end uses a multi-stream approach to incorporate, in addition to MFCCs, auditory-based phonetic distinctive cues. These features are combined in order to limit the impact of the speech signal degradations due to interfering noise. Some of many advantages of using GAs include the possibility to improve the robustness without modifying the recognition system models and without estimating environment parameters, such as the noise variance and/or stream weights. The co-existence, in two streams, of the two types of front-end parameters (MFCCs and distinctive cues) is also managed by the GA. The evaluation is carried out by using a noisy version of the TIMIT corpus.

Keywords Hidden Markov Models • Genetic Algorithms • KLT • Variance of reconstruction error • Acoustic indicative features • TIMIT corpus

8.1 Expected Advantages

Evolutionary computation is a class of soft computing derived from biological concepts and evolution theory. Systems using evolutionary principles model a problem in such a way that the solution is optimized and it keeps improving over time. Evolutionary Algorithms (EAs), that are the most important sub-fields of evolutionary computing, can be considered as heuristic search techniques based on the principles of natural selection. EAs involve various populations of solutions that undergo transformations by using genetic operators that help to converge to the best solution. Selection plays an important role in the evolutionary-based process. In most applications, the determination of the selection function requires an explicit evaluative function which should be interpretable and meaningful in terms of performance, and it is usually prepared by the human designer [45].

Evolutionary Algorithms include genetic algorithms, evolution strategies, evolutionary programming and genetic programming. They have been successfully applied to various domains including machine learning, optimization, bioinformatics

S.-A. Selouani, *Speech Processing and Soft Computing*, SpringerBriefs in Electrical and Computer Engineering, DOI 10.1007/978-1-4419-9685-5_8,
© Springer Science+Business Media, LLC 2011

and social systems [37]. However, their use in speech recognition is still very limited probably because it is very difficult to design human evaluation explicit functions for speech recognition systems since we cannot know in advance what utterance will be spoken. Among the EAs, Genetic Algorithms (GAs) have become an increasingly appreciated and well-understood paradigm beyond the soft computing community [127]. In the field of speech recognition robustness, investigating innovative strategies becomes essential to overcome the drawbacks of classical approaches. For this purpose, GAs can constitute robust solutions as they have demonstrated their power to find optimal solutions in complex problems. The main advantage of GAs is their relative simplicity.

Let's consider $P(t)$ as a population of individuals at time t, $\Phi(.)$ as a random operator and $\Psi(.)$ as the individuals' selection function. The procedure underlying the GA which leads to the population of the next generation can be formalized by the following equation:

$$P(t + 1) = \Phi(\Psi(P(t))) \tag{8.1}$$

It should be mentioned that in contrast to other formal methods, the performance of GA is not impacted by the representation of the population. Thanks to the $\Psi(.)$ function, GAs also offer the possibility to incorporate prior(human) knowledge of the problem, thus yielding to better solution accuracy. GAs can also be combined to other soft computing and optimization techniques (e.g. tuning neural networks structure [26]) by modifying the $\Phi(.)$ operator.

In contrast to many other optimization methods, parallelization of the GAs is possible. In some applications, parallel implementations are necessary to reach high-quality solutions in a reasonable time span [43]. Another important advantage is the ability of GAs to be robust towards dynamic changes. They also do not require a complete restart of the process when an environment change occurs [3]. In previous work [131], we have demonstrated the efficiency of a solution dealing with genetic optimization of NN-based digit recognizer. To improve the robustness of speech recognition, we investigate the hybridization of GAs with the KLT subspace decomposition using the VRE method (described in Chapter 4). This approach uses local search information and mechanisms to achieve complementarity between the genetic algorithm optimization and KLT-VRE subspace decomposition.

8.2 Problem Statement

The principle of GAs consists of manipulating a population of solutions and implementing a 'survival of the fittest individual' strategy to find the best solution. Simulating generations of populations, the fittest individuals of any population are encouraged to reproduce and survive across successive generations to improve both the overall and individual performance. In some GA implementations a proportion of less performing individuals can survive and also reproduce. A more complete presentation of GAs can be found in [129].

Limiting the drop in speech recognition performance in the context of acoustic environment changes remains one of the most challenging issues of speech recognition in practical applications. This misperformance is due to the unpredictability of adverse conditions that create mismatches between the training data and the test data used by the recognizers. New strategies are needed to make the ASR not only robust but also capable of self-adaptation to variable acoustic conditions. In the ideal situation, the ASR is expected to be capable of perceiving the environment changes and to adapt key features (models, front-end process, voice activity detector parameters,...) to the new context. In their pioneering work, Akbacak and Hansen proposed an original framework called *Environmental Sniffing* that aims to perform smart tracking of environmental conditions and to guide the ASR engine to the best local solution adapted to each environmental condition [7]. In this chapter, we investigate the use of GAs in order to optimize the front-end processing of ASR systems. The front-end features are composed of MFCCs and auditory-based phonetic distinctive cues. The expected advantages of using GAs is that the ASR robustness can be improved without modifying the recognition models and without determining the noise variance.

8.3   Multi-Stream Statistical Framework

HMMs constitute the most successful approach developed for modeling the statistical variations of speech in an ASR system. Each individual phone (or word) is represented by an HMM. In large-vocabulary recognition systems, HMMs usually represent subword units, either context-independent or context-dependent, to limit the amount of training data and storage required for modeling words. Most recognizers typically use left-to-right HMMs, which consist of an arbitrary number of states N. The output distribution associated with each state is dependent on one or more statistically independent streams. To integrate the proposed features in the input vector, we merged different sources of information about the speech signal extracted from both the cepstral analysis and Caelen's auditory-based analysis. The multi-stream paradigm is used for modeling the statistical variations of each information source (stream) in an HMM-based ASR system [135]. In this paradigm, an observation sequence $\mathbf{O}$ composed of S input streams, $\mathbf{O}_s$ possibly of different lengths, is assumed as representative of the utterance to be recognized, and the probability of the composite input vector $\mathbf{O}_t$ at a time t in state j can be written as follows:

$$b_j(\mathbf{O}_t) = \prod_{s=1}^{S} [b_{js}(\mathbf{O}_{st})]^{\gamma_s} \tag{8.2}$$

where $\mathbf{O}_{st}$ is the input observation vector in stream s at time t and γ_s is the stream weight. Each individual stream probability $b_{js}(\mathbf{O}_{st})$ is represented by the most common choice of distribution, *the multivariate mixture Gaussian*:

$$b_{js}(\mathbf{O}_{st}) = \sum_{m=1}^{M} c_{jsm} \, \mathcal{N}(\mathbf{O}_{st}; \mu_{jsm}, \Sigma_{jsm}) \tag{8.3}$$

where M is the number of mixture components in stream s, c_{jsm} is the weight of each mixture component of state j in each mixture of each stream and $\mathcal{N}(\mathbf{O}; \mu, \Sigma)$ denotes a multivariate Gaussian of mean μ and covariance Σ and can be written as:

$$\mathcal{N}(\mathbf{O}; \mu, \Sigma) = \frac{1}{\sqrt{(2\pi)^n |\Sigma|}} \exp^{-\frac{1}{2}(\mathbf{O}-\mu)'\Sigma^{-1}(\mathbf{O}-\mu)} \tag{8.4}$$

In the multi-stream HMM the fusion is assumed to be performed by a single global likelihood probability. The observations are assumed to be generated by each HMM stream with identical topologies and modeled as mixtures of Gaussian densities. In the application presented in this chapter, three streams are considered: MFCCS, the MFCC derivatives and the Caelen Distinctive Cues (CDCs) described in Section 6.3.2.

8.4 Hybrid KLT-VRE-GA-based Front-End Optimization

The principle of the signal subspace techniques consists of constructing an orthonormal set of axes forming a representational basis that projects towards the direction of maximum variability. Applied in the context of noise reduction, these axes permit us to decompose the space of the noisy signal into a signal-plus-noise subspace and a noise subspace. As seen previously, the enhancement is performed by removing the noise subspace and estimating the clean signal from the remaining signal space.

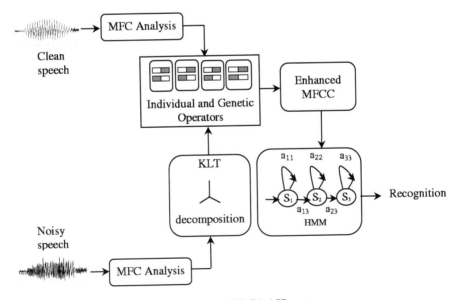

Fig. 8.1 General overview of the hybrid KLT-VRE-GA ASR system.

In Chapter 5, we have described the framework of evolutionary subspace filtering. This framework is extended in this chapter to the ASR robustness by using GAs to optimize the subspace decomposition of the multi-stream vector. By using GAs, no empirical or *a priori* knowledge is needed at the beginning of the evolution process. As illustrated in Figure 8.1, a mapping operator using a Mel-frequency subspace decomposition and GAs is performed. This evolutionary eigendomain transformation attempts to achieve an adaptation of ASR systems by tracking the best KLT reconstruction after removing the noise subspace using the VRE method.

8.5 Evolutionary Subspace Decomposition using Variance of Reconstruction Error

As shown in Chapter 4, the subspace filtering provides the clean speech estimate, $\hat{s} = Hx$, where $H = UGU^T$ is the enhancement filter containing the weighting factors applied on the eigenvalues of the noisy speech x corrupted by the noise n. In the evolutionary eigendecomposition, the H matrix is replaced by $H_{gen} = UG_{gen}U^T$. The diagonal matrix G_{gen} contains the weighting factors optimized by the genetic operators. Therefore, to improve the ASR performance, the task prior to recognition is finding an estimate for s along the direction ξ_j to best correct the noise effect by using the VRE technique. The number of optimal principal components (PCs) is obtained by achieving the minimum reconstruction error. The reconstruction is performed by using eigenvectors weighted by the optimal factors of the G_{gen} matrix. These factors will constitute the individuals of a given population in the GA process. The mechanism of determining the optimal PCs is performed on many populations in order to achieve the best reconstruction over a wide range of solutions. As specified in Chapter 4, the variance of the reconstruction error in all directions and dimensions can be calculated by:

$$VRE(l) = \sum_{j=1}^{N} \frac{u_j(l)}{\xi_j^T R \xi_j}. \tag{8.5}$$

The *VRE* is calculated by considering the variances u_j (corresponding to the eigenvalues) in all directions and using the R_{xx} the autocorrelation matrix of the noisy signal. This *VRE* has to be minimized to obtain the best reconstruction and will be included in the GA process as an objective function.

8.5.1 Individuals' Representation and Initialization

Any application based on GAs requires the choice of gene representation to describe each individual in the population. In our case the genes are the components of

H_{gen} matrix elements. The real-valued representation is suitable and is expected to obtain more consistent results across replications. An alphabet of floating point numbers has values ranging within the upper and lower bounds of $+1.0$ and -1.0 respectively. This representation is closer to the real representation of the weight factors, and will facilitate the interpretation of optimization results.

To start the evolution process, a pool containing a population of individuals representing the weight factors, g_i is created. Ideally, this pool is initialized with a zero-knowledge assumption by using a population of completely random values of weights. Another approach (guided) consists of performing a first KLT subspace decomposition and using the principal components obtained to constitute the first population pool. Regardless to the initialization method (random or guided), these individuals evolve through many generations in the pool where genetic operators are applied. Some of these individuals are selected to survive and to reproduce according to their performance and other considerations that are taken to insure a good coverage of the solution space. The individuals' performance is evaluated through the use of an objective function.

8.5.2   Selection Function

Various methods exist for the selection of individuals to produce successive generations [127]. Most approaches are based on the assignment of a probability of selection, $Prob_j$ to each individual, j, according to its performance. To perform the selection of the weight factors in the pool, we use a new variant of the normalized geometric ranking method originally proposed in [65]. The probability of selection $Prob_j$ is given by:

$$Prob_j = \frac{q(1-q)^{s-1}}{1-(1-q)^{Pop}},$$ (8.6)

where q is the probability of selecting the best individual, s is the rank of the individual (1 is the rank of the best individual), and Pop is the population size. All solutions are sorted and ranked. To insure the solution diversity, a mechanism giving more chance of selection to a small proportion of worse performing individuals is applied. This mechanism is formalized as follows:

$$Prob'_j = \begin{cases} Prob_j & \text{for } (s = 1, ..., Pop/2) \\ Prob_{j+Pop/2-k} & \text{for } (s = Pop/2, ..., Pop) \text{ and } (k = 0, ..., Pop/2) \end{cases}$$ (8.7)

In this selection method, the pool is divided into two groups. The first group contains individuals that perform better than the elements of the second group. The regular ranking, according to the fitness value, is applied in this first group. In the second group, the ranking is inverted in order to ensure and maintain the population diversity by giving a chance to some of the less performing individuals to be selected. This diversity allows the GA to perform fruitful exploration by expanding the search space of solutions. Therefore, in the selection process, $Prob_j$ is used to select individuals.

8.5.3 Objective Function

The performance of any individual k is measured by an objective function $\mathcal{F}(k)$ also called the fitness function. In GAs two types of objective functions can be considered. The first type is the best fitness, where the retained solution corresponds to the individual having the best performance. The second type is average fitness, which provides the solution corresponding to the average of the best individuals after a certain number of runs. The objective (fitness) function is defined in terms of a performance measure and gives a quantifiable way to rank the solutions from good to bad. In the KLT-VRE-GA framework, the best solution corresponds to the individual minimizing the VRE function given in Equation 8.5. Therefore, the objective function to minimize can be written as follows:

$$\mathcal{F}(k) = \min[VRE(k)]. \tag{8.8}$$

8.5.4 Genetic Operators and Termination Criterion

Genetic operators use the objective and selection functions to apply some modifications on selected individuals to produce offspring of the next generation. This process provides new possibilities in the solution space by combining the fittest chromosomes and passing superior genes to the next generation. There are numerous implementations of genetic operators. For instance, there are dozens of possible crossover and mutation operators that have been developed in recent years.

A heuristic crossover generating a random number v from a uniform distribution and doing an exchange of the parents' genes (X and Y) on the offspring genes (X and Y) is used in this application. The main characteristic of this type of crossover is that it utilizes the fitness information. Offspring are created using the following equation:

$$X' = X + v(X - Y)$$
$$Y' = X, \tag{8.9}$$

where X is assumed to perform better than Y, in terms of objective function. Heuristic crossover introduces a *feasibility* function Fs, defined by:

$$Fs(X') = \begin{cases} 1 \text{ if } a_i \leq x'_i \leq b_i \quad \forall i \\ 0 \text{ otherwise,} \end{cases}$$

where x_i' are the components in N-dimensional space, of X' with $i=1,...,N$. The Fs function controls the generation of a new solution using Equation 8.9. In fact, when $Fs(X')$ equals 0, a new random number v is generated in order to create the offspring.

Mutation operators lead to small random changes of the individual components in an attempt to explore more regions of the solution space [29]. The principle of a non-uniform mutation used here consists of randomly selecting one component x_k of an individual and setting it equal to a non-uniform random number, otherwise, the original values of components are maintained. The new component, x_k', is given by:

$$x_k' = \begin{cases} x_k + (b_k - x_k) f(Gen) & \text{if } u_1 < 0.5 \\ x_k - (a_k + x_k) f(Gen) & \text{if } u_1 \geq 0.5 \end{cases} \qquad (8.10)$$

where the function $f(Gen)$ is given by:

$$f(Gen) = \left(u_2 \left(1 - \frac{Gen}{Gen_{max}} \right) \right)^t, \qquad (8.11)$$

u_1, u_2 are uniform random numbers selected within $(0,1)$, t is a shape parameter, Gen the current generation and Gen_{max} the maximum number of generations. We have shown in [123] that the use of the heuristic crossover and the non-uniform mutation [65] are suited for the reduction of additive noise.

When a certain number of predetermined generations is reached, the evolution process is terminated. The fittest individual, which corresponds to the best set of weights or optimal axes, is then used to project the noisy data. Then, the "genetically modified" MFCCs and CDCs are used as enhanced features for the testing phase of the recognition process.

8.6   Experiments and Results

8.6.1   Speech Material

The TIMIT corpus [133] is used to evaluate the KLT-VRE-GA approach. Timit contains broadband recordings of 6300 sentences: 10 sentences spoken by each of 630 speakers from 8 major dialect regions of the United States, each reading 10 phonetically rich sentences. All train subsets of the TIMIT database are used to train the models and the test sub-directories are used to evaluate the recognition systems.

Table 8.1 Genetic
parameters used in the
application.

Parameter	Parameter Value
Number of generations	500
Population size	200
Probability of selecting the best g_i	0.10
Heuristic crossover rate	0.35
Multi-Non-Uniform Mutation rate	0.05
Number of runs	70

8.6.2 Recognition Platform

The HTK HMM-based speech recognition system described in [66] has been used
throughout all experiments. The HTK toolkit was designed to implement HMMs
with any numbers of state and mixture components. It also allows the creation of
complex model topologies to suit a variety of speech recognition applications. All
the tests are performed using N-mixture ($N = 1, 2, 4, 8$) Gaussian HMMs with
tri-phone models.

8.6.3 Tests & Results

To simulate a noisy environment, various noises are added artificially to the clean
speech at different SNR levels varying from 16 dB to -4 dB. The reference models
are created using clean speech. Four different sets of experiments are designed. The
first set concerns the baseline system in which 12 MFCCs are calculated over a 30-
msec Hamming window. The normalized log energy is replaced by the mid-external
energy of the ear extracted by means of the Caelen's model. The dynamic features
that are the first and second derivatives of MFCCs are also included. Furthermore,
in order to compare the KLT-VRE-GA system with a recognizer using a well-
established noise-reduction technique, the mean normalization of MFCCs (CMN)
is applied to the 12-dimensional MFCC vector. The second set involves the well-
known state-of-the-art eigen-decomposition method, the one based on the KLT
applied in the Mel-frequency space (c.f. Chapter 3). The third set of tests carries out
speech recognition using the evolutionary-based eigen-decomposition (KLT-VRE-
GA) method. In the last set of tests, the static vector composed of the 36 MFCCs
and their derivatives is expanded by adding the 7 CDCs and the mid external energy
of the ear, to form a 44-dimensional vector upon which the baseline multi-stream
HMMs were trained.

The values of the GA parameters used in our experiments are given in Table 8.1.
To realize the compromise between speed and accuracy, a population of 200 individuals is generated for each axis. The objective function stabilizes (no improvement is
noticed) after approximately 300 generations. In order to insure the convergence in
all situations, the maximum number of generations was finally fixed at 500.

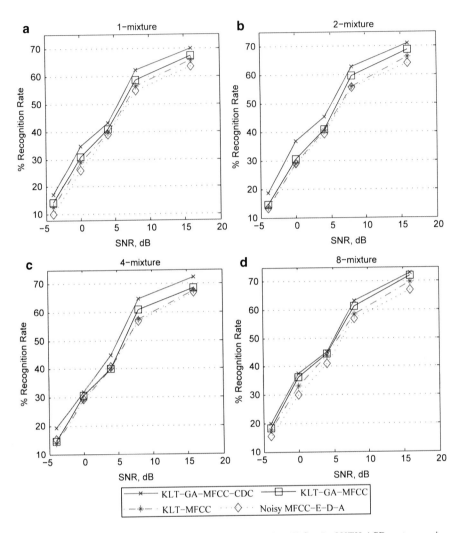

Fig. 8.2 Comparisons of the percentage of word recognition ($\%C_{wrd}$) of HTK ASR systems using N-mixture ($N = 1, 2, 4, 8$) triphones and TIMIT database corrupted by additive car noise: the first ASR system uses MFCCs and their first and second derivatives with mean normalization. The second system uses the KLT normalized MFCCs using the VRE objective function, the third includes the KLT-GA-based front-end applied to MFCCs, and finally the fourth ASR system applies the KLT-GA to the CDCs and MN-MFCCs.

As shown in Figure 8.2, the system including the KLT-GA based front-end achieves higher accuracies compared to the baseline system dealing with noisy normalized MFCCs. This improvement is observed for all SNR values and varies within a range of 3% and 8%. The KLT-GA-CDC system which combines the MFCCs, their first and second derivatives and CDCs outperform the other systems for all values of SNR.

8.7 Summary

This chapter has presented a promising approach advocating the usefulness of evolutionary-based subspace filtering to complement conventional ASR systems meant to tackle the challenge of noise robustness. In fact, the combined effects of subspace filtering optimized by GAs and knowledge gained from measuring the auditory physiological responses to speech stimuli may provide more robustness to speech recognition. It should be noted that such a soft computing technique is less complex than many other robust techniques that need to either model or compensate for noise. Many other directions remain open. The research on acoustic-phonetic features in ASR should benefit more from the knowledge related to the auditory system. A promising way is to modify the basic preprocessing technique to integrate phonetic knowledge directly. Distinctive cues can be learned by an arbitrary function approximator such as an artificial neural network (ANNs), yet another soft-computing technique. The training of ANNs on acoustic distinctive feature labels will permit us to gain a more effective representation of the acoustic-phonetic mapping function. Using this approach avoids the noise estimation process that requires a speech/non-speech pre-classification, which could not be accurate for low SNRs.

Chapter 9
Speaker Adaptation Using Evolutionary-based Approach

Abstract Speaker adaptation is one of the most important areas of speech recognition technology which continues to attract the interest of researchers. Current speaker adaptation methods seek to achieve both fast and unsupervised adaptation by using a small amount of data. In the present chapter, we present an approach that aims at investigating more solutions while simplifying the adaptation process. In this approach, a single global transformation set of parameters is optimized by genetic algorithms using a discriminative objective function. The goal is to achieve accurate speaker adaptation, whatever the amount of available adaptive data.

Keywords Speaker adaptation • Genetic Algorithms • MLLR • Eigendecomposition • Discriminative adaptation • ARPA-RM corpus

9.1 Speaker Adaptation Approaches

Current speech recognition systems achieve a very high recognition rate in a speaker-dependent context (SD), but their performance often degrades when mismatches between training and testing conditions are introduced by new speakers. To cope with these mismatches, a simple retraining using the new speaker data can be performed. However, in real-life applications it is very difficult to acquire a large amount of training data from a new test speaker. There is a wide variety of speaker adaptation techniques that are applied to the continuous density hidden Markov models. These methods fall into three categories based on linear transforms of HMMs' parameters such as MLLR [84]; speaker space decomposition methods such as eigenvoices [80,97]; and MAP adaptation [83].

In linear transform methods, a global transformation matrix is estimated in order to create a general model which better matches a particular target condition generated by the new speaker. To perform the adaptation on a small amount of data, a regression-tree-based classification is performed. The MLLR which is the most popular linear transform technique calculates a general regression transformation for each class, using data pooled within each class. However, as mentioned in [95], transformation-based adaptation techniques suffer from two principal drawbacks. The first drawback is related to the fact that the type of the transformation function is fixed in advance to simplify the mathematical formalism. The second drawback is their bad asymptotic properties. This means that these techniques may not achieve the level of accuracy obtained with speaker-dependent systems even if the adaptation data quantity increases largely. The MAP-based techniques have better asymptotic properties but require more adaptation data compared to linear transform methods. Over the last few years, eigenvoice methods have become the backbone of most speaker adaptation methods. Eigen voice modeling performs unsupervised and fast speaker adaptation through the use of eigen-decomposition, where the principal component analysis is used to project utterances of unknown speakers onto the orthonormal basis leading to SD eigen coefficients.

The straightforward approach estimates the linear transform parameters that could be the mean and/or the variance of the speaker-independent (SI) HMMs. These parameters are used to perform the retraining by applying the maximum likelihood (ML) criterion to adjust the SI acoustic models so that they better fit the characteristics of a new speaker. Another recent and widely employed alternative approach consists of using discriminative linear transforms (DLT) to construct more accurate speaker adaptive speech recognition systems. Well-known discriminative criteria include maximum mutual information (MMI), minimum classification error (MCE), and minimum phone error (MPE) training. In [144], the MPE criterion is adopted for DLT estimation. Uebel and Woodland in [137] performed an interpolation of ML and MMI training criteria to estimate the DLT. In [103], Povey *et al.* studied the incorporation of the MAP algorithm into MMI and MPE for task and gender adaptation.

Many extensions have been proposed to improve the basic schemes of conventional and discriminative speaker adaptation, resulting in a wide range of hybrid approaches. However, very few methods explicitly include soft computing in their core scheme. In [81], GAs have been used to enrich the set of SD systems generated by the eigen-decomposition. Here, we propose a speaker adaptation technique based on the determination of a single global transformation set of parameters optimized by genetic algorithms using a discriminative objective function. Through the use of this evolutionary-based method, we expect to improve the accuracy of eigen-MLLR techniques.

9.2   MPE-based Discriminative Linear Transforms for Speaker Adaptation

Discriminative training algorithms, such as the MMI and MPE, have been successfully applied in large vocabulary speech recognition [104, 147] and speaker adaptation tasks [145]. The main characteristic of these algorithms is that they

Fig. 9.1 Overview of the the MPE-based discriminative training for speaker adaptation.

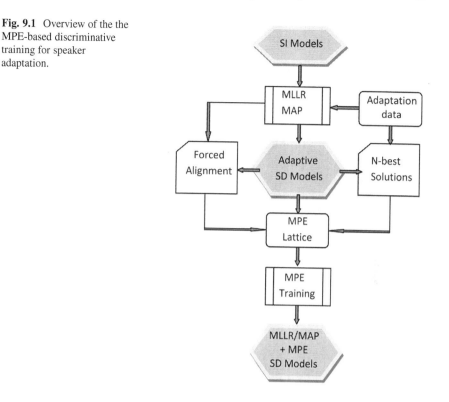

consider not only the correct transcription of the training utterance, but also the competing hypotheses that are obtained by performing the recognition step.

In order to facilitate the inclusion of an evolutionary-based optimization, a baseline system is constructed by performing the MPE training to improve the SD acoustic models obtained by the MLLR. This system is depicted by the block diagram given in Figure 9.1. The SI model is first adjusted by MLLR using limited speaker-specific data. Then, the adapted SD model is updated by a MPE-based discriminative training. The numerator lattice is obtained through the alignment process on the transcriptions of the adaptation data. The denominator lattice is approximated with the N-best phone hypotheses after performing the recognition process on the adaptation data.

In the approach presented here, an MPE discriminative training is performed by using speaker-specific data. Many studies have demonstrated that MPE training outperforms MMI training [144]. Actually, MPE focuses on correctable errors in the training data rather than outliers which may reduce the effectiveness of MMI training. The MPE-based method consists of using a weak-sense auxiliary function in HMM to re-estimate the mean $\tilde{\mu}_{km}$ of mixture component m of state k of a new adapted model. This re-estimation is done as follows:

$$\tilde{\mu}_{km} = \frac{[\theta_{km}^{num}(O) - \theta_{km}^{den}(O)] + D_{km}\hat{\mu}_{km}}{[\gamma_{km}^{num} - \gamma_{km}^{den}] + D_{km}}, \tag{9.1}$$

where $\theta_{km}^{num}(O)$ and $\theta_{km}^{den}(O)$ are respectively the numerator and denominator sum of observation data weighted by the occupation probability for mixture m of state k; D_{km} is the Gaussian-specific smoothing constant; γ_{km}^{num} and γ_{km}^{den} are respectively the numerator occupation probabilities and the denominator occupation probabilities summed over time.

State-of-the-art techniques show that two different forms of discriminative speaker adaptation techniques (DSAT) are being used [112]. The first technique is based on ML speaker-specific transforms, and its commonly used variant is the MLLR-based DSAT. In this approach, both ML-based and discriminative training are used. The MLLR-based adaptation is initially performed to produce a set of speaker-specific MLLR transforms. These transforms are then used to carry out the subsequent updates by using the MPE discriminative criterion. As stated by Raut et al. in [112] the use of ML-based speaker-specific transforms leads to more robustness to errors in the supervision hypothesis. The second approach is based on discriminatively estimated transforms that are referred to as DLTs. In these DLT-based methods, both of the transforms and the HMMs are estimated by using the MPE discriminative criterion. This yields a set of speaker-specific DLTs that are used for recognition. For the experiments presented in this chapter, only the MLLR-based DSAT is used for comparison purposes.

9.3 Evolutionary Linear Transformation Paradigm

Genetic algorithms have been successfully integrated in the framework of speaker adaptation of acoustic models [124]. One of the approaches consists of using the genetic algorithm to enrich the set of speaker-dependent systems employed by the eigenvoices [81]. In this latter work, the best results are obtained when the genetic algorithms are combined with the eigen decomposition. Since the eigen decomposition provides the weights of eigenvoices by using the EM algorithm, it can only find a local solution. In the GA-MPE-MLLR method presented here, the eigen-decomposition is avoided and the MPE criterion is used as an objective function. The MPE-based training has proven to be very effective in the generalization from training to test data, when compared with the conventional maximum likelihood approach. The motivation for an evolutionary-based discriminative transform is based on the fact that DLTs were initially developed to correctly discriminate the recognition hypotheses for the best recognition results rather than just to match the model distributions.

In the GA-MPE-MLLR method, the mean transformation matrix (obtained by MLLR) provides the individuals involved in the evolutionary process. As shown in Section 6.4.3, μ_k is the baseline mean vector and $\hat{\mu}_k$ is the adapted mean

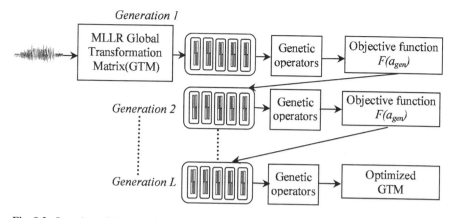

Fig. 9.2 Overview of the evolutionary-based non-native speaker adaptation system.

vector for an HMM state k. The relationship between these two vectors is given by: $\hat{\mu}_k = \mathbf{A}_k \xi_k$ where $\mathbf{A}_k$ is the $d \times (d+1)$ transformation matrix and $\xi_k = [1, \mu_{k1}, \mu_{k2}, ..., \mu_{kd}]^t$ is the extended mean vector. The $\mathbf{A}_k$ matrix will contain weighting factors that represent the individuals in an evolution process. These individuals evolve through many generations in a pool where genetic operators such as mutation and crossover are performed [29]. Some of these individuals are selected to reproduce according to their performance. The individuals evaluation is performed through the use of the objective function. The evolution process is terminated when no improvement of objective function is observed. When the fittest individuals are obtained (the global optimized matrix $\mathbf{A}_{\mathbf{gen}}$), they are used in the test phase to adapt the data of new speakers. It is important to note that we do not need to determine the regression classes, since the optimization process is driven by a performance maximization whatever the amount of available adaptive data. The GA-based adaptation process is illustrated by Figure 9.2. For any GA, a chromosome representation is needed to describe each individual in the population. The representation scheme determines how the problem is structured in the GA and also determines the genetic operators that are used. GA-MPE-MLLR involves genes that are represented by the components of $\mathbf{A}_{\mathbf{gen}}$ matrix elements.

9.3.1 Population Initialization

The first step to start the GA-MPE optimization is to define the initial population. This initial population is created by 'cloning' the elements of a global $\mathbf{A}$ matrix issued from a first and single MLLR pass. This procedure consists of duplicating the a_i elements of $\mathbf{A}$ (given initially by Equation 6.28) to constitute the initial pool with a predetermined number of individuals. Hence, the pool will contain a_i^v individuals where v refers to an individual in the pool and it varies from 1 to *PopSize* (population size). With this procedure, we expect to exploit the efficiency of GAs to explore the entire search space, and to avoid a local optimal solution. The useful representation of individuals involves genes or variables from an alphabet of floating point numbers with values varying within lower and upper bounds (b_1, b_2).

9.3.2 Objective Function

The optimization of the global transformation matrix requires finding the fittest individuals representing the column vectors, denoted $a_i^v \in S$, where S is the search space, so that a certain quality criterion is satisfied. In our case, this criterion states that objective function $\mathcal{F} : S \rightarrow \mathcal{R}$ is maximized. Therefore, $a_{i_{gen}}$ is the solution that satisfies:

$$a_{i_{gen}} \in S \mid \mathcal{F}(a_{i_{gen}}) \geq \mathcal{F}(a_i^v) \qquad \forall a_i^v \in S. \tag{9.2}$$

In the method we propose, the objective function (fitness) is defined in such a way that the newly genetically optimized parameters are guaranteed to increase the phone accuracy of adaptation data. For this purpose, we used the minimum phone error criterion utilizing phone lattices. The standard function reflecting the MPE criterion involves competing hypotheses represented as word lattices, in which phone boundaries are marked in each word to constrain the search during statistical estimation of an HMM model λ. For a specific model, this function is defined as:

$$F_{MPE}(\lambda) = \sum_{u=1}^{U} \sum_{s} P_\rho(s|O_u, \lambda) \sum_{q \in s} PhAcc(q), \tag{9.3}$$

where $P_l(s|O_u, \lambda)$ is the posterior probability of hypothesis s for utterance u given observation O_u, current model λ and acoustic scale ρ. $\sum_{q \in s} PhAcc(q)$, is the sum of phone accuracy measure of all phone hypotheses. The objective function used in the GA-MPE to evaluate a given individual a_i^v, considers the overall phone accuracy and then it is defined as:

$$\mathcal{F}(a_i^v) = \sum_{\lambda} F_{MPE}(\lambda). \tag{9.4}$$

Objective function is normalized to unity. Figure 9.3 plots variations of the best individual $\mathcal{F}(a_{i_{gen}})$ with respect to the number of generations, in the case of totally random and first step MLLR initializations of the population.

9.3.3 Selection Function

Since the offspring population is larger than the parent population, a mechanism has to be implemented in order to determine the individuals that will comply with the new parent population. The selection mechanism chooses the fittest individuals

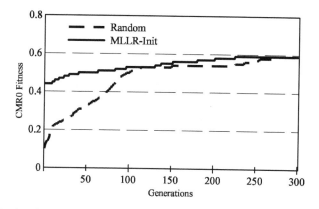

Fig. 9.3 Objective function variations with random and basic MLLR initializations of the population.

of the population and allows them to reproduce, while removing the remaining individuals. The selection of individuals to produce successive generations is based on the assignment of a probability of selection, P_v to each individual, v, according to its fitness value. In the 'roulette wheel' method [49], the probability P_v is calculated as follows:

$$P_v = \frac{\mathcal{F}(a_i^v)}{\sum_{k=1}^{PopSize} \mathcal{F}(a_i^k)} \qquad (9.5)$$

where $\mathcal{F}(a_i^k)$ equals the value of objective function of individual k and *PopSize* is the population size in a given generation. In the 'roulette wheel' variant implemented in GA-MPE, we introduced a dose of an *elitist* selection by incorporating in the new pool, the top two parents of previous populations in order to replace the two offspring individuals having the lowest fitness.

9.3.4   Recombination

Recombination allows for the creation of new individuals (offspring) using individuals selected from the previous generation (parents). In the GA-MPE method, a combination of the conventional arithmetic crossover and guided crossover is used as a recombination operator. In the first step, this method selects an individual as a first candidate ($cand_1$). A second candidate is then selected according to a quantity of what is called the mutual fitness $MF(X, cand_1)$ [111], where a choice for X as a second candidate is made if it maximizes the mutual fitness with the first candidate. The general computation of the mutual fitness is given by:

$$MF(A, B) = \frac{[\mathcal{F}(A) - \mathcal{F}(B)]^2}{Distance(A, B)^2} \qquad (9.6)$$

The parents $cand_1$ and $cand_2$ are now selected and the convex combination can be applied according to the following equations :

$$\begin{cases} mix = (1 + 2 * \beta) * rand - \beta \\ x' = mix * cand_1 + (1 - mix) * cand_2 \\ y' = (1 - mix) * cand_1 + mix * cand_2, \end{cases} \qquad (9.7)$$

where $rand$ is a Gaussian random value. If β is set to 0, the resulting crossover is a simple crossover. If β is set to a positive value this may increase the diversity of the individuals of the population, and may allow the offspring to explore the solution space beyond the domain investigated by their parents.

9.3.5   Mutation

Mutation operators tend to make small random changes on the individual components in order to increase the diversity of the population. Mutation consists of randomly selecting one gene x of an individual a_i^X and slightly perturbating it. In GA-MPE, the offspring mutant gene, x'', is given by:

$$x'' = x + \mathcal{N}_k(0, 1) \qquad (9.8)$$

where $\mathcal{N}_k(0, 1)$ denotes a random variable of normal distribution with zero mean and standard deviation 1 which is to be sampled for each component individually. The Gaussian-based alteration on the selected offspring individuals allows the extension of the search space and theoretically improves the ability to deal with new speaker related conditions.

9.3.6   Termination

The evolution process is terminated when a number of maximum generations is reached. No improvement of the objective function is observed beyond a certain number of generations. It is also important to note that as expected, the single class MLLR initialization yields a rapid fitness convergence, in contrast to the fully random initialization of the pool. When the fittest individual is obtained, it is used to produce a speaker-specific system from an (SI) HMM set.

9.4 Experiments

9.4.1 *Resources and Tools*

The TIMIT and ARPA-RM databases were used to evaluate the MPE-GA-MLLR technique. The Train subset of TIMIT is used for the training while a speaker dependent subset of ARPA-RM consisting of 47 sentences of ARPA-RM uttered by 6 speakers is used for the test. The HTK toolkit implementing HMM-based speech recognition system is used throughout all experiments. The adaptation was performed in an unsupervised mode. The testing adaptation is performed with an enrollment set of 10 sentences. The acoustical analysis consists of 12 MFCCs which were calculated on a 30-msec Hamming window. The normalized log energy, the first and second derivatives are added to the 12 MFCCs to form a 39-dimensional vector. All tests are performed using 8-mixture Gaussian HMMs with tri-phone models.

9.4.2 *Genetic Algorithm Parameters*

To control the run behaviour of a genetic algorithm, a number of parameter values must be defined. The initial population is composed of 200 individuals and is created by duplicating the elements of global transform matrix obtained after the first and single regression class MLLR. The genetic algorithm is halted after 350 generations. The percentage of crossover rate and mutation rate are fixed at 35% and 8%, respectively. The number of total runs is fixed at 60. The GA-MPE-MLLR system uses a global transform where all mixture components are tied to a single regression class.

9.4.3 *Result Discussion*

Table 9.1 summarizes the word recognition rates obtained for the 6 speakers using four systems: the baseline HMMs-based system without any adaptation (unadapted) and using the ML criterion for recognition; the conventional MLLR using the ML criterion; the MLLR using a discriminative transformation (MLLR-DSAT) described in Section 9.2; and the system integrating the evolutionary subspace approach using the MPE criterion (GA-MPE-MLLR). The GA-MPE-MLLR system achieves an improvement in the accuracy of word recognition rate reaching 8% compared to the baseline unadapted system and more than 3% compared to conventional MLLR. We have tested the fully random initialization of the population and the one using individuals cloned from MLLR global transformation matrix components. In

Table 9.1 Comparisons of the percentage of word recognition ($\%C_{wrd}$) of HMM-based CSR systems for selected data from the ARPA-RM used for adaptation and test, while the TIMIT $dr1$ and $dr2$ subsets were used for training.

Speaker	CMR0	DAS1	DMS0	DTB0	ERS0	JWS0
Unadapted	76.46	75.14	78.65	80.24	76.97	77.86
MLLR-ML	78.34	79.87	80.74	84.92	83.56	80.85
MLLR-DSAT	78.15	77.49	80.82	84.20	83.78	81.73
GA-MPE-MLLR	**79.71**	**81.30**	**82.14**	**86.26**	**84.75**	**83.92**

both cases, the final performance is the same. However, the adaptation is reached rapidly (160 generations) with a MLLR-based initialization.

9.5 Summary

The most popular approaches to speaker adaptation are based on linear transforms because they are considered more robust and use less adaptation data than the other approaches. This chapter has presented a framework demonstrating the suitability of the soft-computing technique based on genetic algorithms to improve unsupervised speaker adaptation using linear transforms. In fact, experiments have shown that the GA-MPE-MLLR approach outperforms the discriminative and conventional MLLR speaker adaptation technique. The main advantage of using the soft computing based optimization approach is to avoid the regression class process usually done in conventional MLLR. Therefore, the new speaker adaptation performance is not linked to the amount of available adaptive data. Many perspectives are open and may consist of fully automating the set up of genetic parameters. The ultimate objective is to give CSR systems auto-adaptation capabilities to face any acoustic environment changes.

References

1. Abad, A., Pellegrini, T., Trancoso, I., Neto, J., "Context dependent modelling approaches for hybrid speech recognizers," in *Proc. Interspeech*, pp. 2950-2953, 2010.
2. Abolhassani, A., Selouani, S.-A., O'Shaughnessy, D., Harkat, M.F., "Speech Enhancement Using PCA and Variance of the Reconstruction Error Model Identification," in *Proc. Interspeech*, pp. 974-977, Belgium, August 2007.
3. Ajith, A, Nedjah, N., Mourelle, L.D.M., "Evolutionary Computation: from Genetic Algorithms to Genetic Programming," *Studies in Computational Intelligence (SCI)*, Springer-Verlag Berlin Heidelberg, 13, pp. 120, 2006.
4. ALICE, AI Foundation, "Artificial Intelligence Markup Language (AIML)," A.L.I.C.E. AI Foundation Working Draft, 8 August 2005 (rev 008), http://www.alicebot.org/ .
5. Allen, J.B., "How do Humans Process and Recognize Speech?," *IEEE Trans. Acoust., Speech, Signal Process.*, 2(4), 567-577, 1994.
6. Akaike, H., "Information theory and an extension of the maximum likelihood principle," in *Proc. 2nd International Symposium on Information Theory*, Petrov and Caski, Eds., pp 267-281, 1974.
7. Akbacak, M., Hansen, J. H. L., "Environmental Sniffing: Noise Knowledge Estimation for Robust Speech Systems," *IEEE Trans. Acoust., Speech, Signal Process.*, vol. 15, no. 2, pp. 465-477, Feb. 2007.
8. Baker, J. E., "Reducing Bias and Inefficiency in the Selection Algorithm," *in Proc. Second International Conference on Genetic Algorithms and their Application*, pp. 14-21, 1987.
9. Bartkova K., Jouvet D., "Multiple Models for Improved Speech recognition For Non-Native Speakers," *in Proc. 9th Conference of Speech and Computer*, pp. 22-28, St. Petersburg, Russia 2004.
10. Ben Aicha, A., Ben Jebara, S., "Perceptual Musical Noise Reduction using Critical Band Tonality Coefficients and Masking Thresholds," *Interspeech Conf.*, pp. 822-825, Antwerp, Belgium, 2007.
11. Benahmed, Y., Selouani, S.A., O'Shaughnessy, D., "Real-life Speech-Enabled System to Enhance Interaction with RFID Networks in Noisy Environments," in *Proc. IEEE ICASSP*, pp. 1781-1788, May 2011.
12. Benesty, J., Makino, S., Chen, J., "Speech Enhancement," *Springer Series: Signals and Communication Technology*, 406 pages, 2005.
13. Benesty, J., M.M. Sondhi, M.M., Huang, Y., *Springer Handbook of Speech Processing*. Springer-Verlag, Berlin, Germany, 2007.
14. Boll, S.F., "Suppression of acoustic noise in speech using spectral substraction," *IEEE Trans. Acoust., Speech, Signal Process.*, vol. 29, pp. 113-120, 1979.

S.-A. Selouani, *Speech Processing and Soft Computing*, SpringerBriefs in Electrical and Computer Engineering, DOI 10.1007/978-1-4419-9685-5,
© Springer Science+Business Media, LLC 2011

15. Bourlard, H., Morgan, N., *Connectionist Speech Recognition: A Hybrid Approach*. Kluwer Publisher, 1994.
16. Brown, R., *Exploring New Speech Recognition And Synthesis APIs In Windows Vista*. MSDN magazine, January 2006.
17. Cadzow, J., "Signal Enhancement. A Composite Property Mapping Algorithm," *EEE Trans. Acoust., Speech, Signal Process.*, ASSP-36, pp. 49-62, 1988.
18. Caelen, J., Space/Time data-information in the ARIAL project ear model, *Speech Communications*, 4(1), 1985.
19. Carré, R., Descout, R., Eskenazi, M., Mariani, J., and Rossi, M. "French language database: defining, planning and recording a large database," *in Proc. IEEE ICASSP*, pp. 324-327, 1984.
20. Chen, L.-Y., Lee, C.-J., Jang, J.-S. R., "Minimum phone error discriminative training for Mandarin Chinese speaker adaptation," in *Proc. Interspeech*, pp. 1241-1244, 2008.
21. Chen, M.Y., "Acoustic correlates of English and French nasalized vowels," *J. Acoust. Soc. Am.*, vol. 102 (4), pp. 2360-2370, 1997.
22. Ching-Ta, L. "Enhancement of single channel speech using perceptual-decision-directed approach," *Speech communication*, Elsevier, vol. 53(4), pp. 495-507, 2011.
23. Chomsky, N., & Halle, M., *Sound pattern of English*. New York: Harper and Row, 1968.
24. Cohen, I., "Noise Estimation by Minima Controlled Recursive Averaging for Robust Speech Enhancement," *IEEE Signal Processing Letters.*, 9(1), pp. 12-15, 2002.
25. Cohen, I., "Noise Estimation in Adverse Environments: Improved Minima Controlled Recursive Averaging," *IEEE Trans. Acoust., Speech, Signal Process.*, 11(5), pp. 466-475, 2003.
26. Correa, A., Gonzalez, A., Ladino, C., "Genetic Algorithm Optimization for Selecting the Best Architecture of a Multi-Layer Perceptron Neural Network: A Credit Scoring Case," *SAS Global Forum 2011 Data Mining and Text Analytics*, paper 149-2011, 2011.
27. Crochiere, R. E., Tribolet, J. E. and Rabiner, L. R., "An interpretation of the Log Likelihood Ratio as a measure of waveform coder performance," *IEEE Trans. Acoust., Speech, Signal Process.*, vol. ASSP-28, no. 3, 1980.
28. Davis, S., & Mermelstein, P., "Comparison of parametric representation for monosyllabic word recognition in continuously spoken sentences," *IEEE Trans. Acoust., Speech, Signal Process.*, 28(4), 357-366, 1980.
29. Davis, L., *The genetic algorithm handbook*. Ed. New York: Van Nostrand Reinhold, 1991.
30. Delvaux, V., Soquet, A., "Discriminant analysis of nasal vs. oral vowels in French: comparison between different parametric representations," in *Proc. Interspeech*, pp. 647-650, 2001.
31. Delvaux, V., Metens, T., Soquet, A., "French nasal vowels: acoustic and articulatory properties," in Proc. of the 7th International Conference on Spoken Language Processing, Denver, 1, pp. 53-56, 2002.
32. Dempster, A. P., Laird, N. M. and Rubin, D. B., "Maximum likelihood from incomplete data via the EM algorithm," *Journal of Royal Statistical Society*, vol. 39, pp. 1-38, 1977.
33. Dendrinos, M., Bakamidis, S. and Carayannis, G., *Speech enhancement from noise: a regenerative approach. Speech Communication*, vol. 10, no. 1, pp. 45-57, 1991.
34. Deng, L., O'Shaughnessy, D., *Speech processing: a dynamic and optimization-oriented approach*. Marcel Dekker Inc., New York, NY., 2003.
35. Diethorn, E.J., *Subband noise reduction methods for speech enhancement*. Gay, S.L., Benesty, J. (Eds.), Acoustic Signal Processing for Telecommunications, Kluwer Academic, Boston, 2000.
36. O'Shaughnessy, D., *Speech communication: Human and machine*. IEEE Press, 2001.
37. Eiben A. E., and Smith J. E., *Introduction to Evolutionary Computing*. Springer, Natural Computing Series, 2nd printing, ISBN: 978-3-540-40184-1, 2007.
38. Ephraim, Y., Mallah, D., "Speech enhancement using a minimum mean-square error short-time spectral amplitude estimation," *IEEE Trans. Acoust., Speech, Signal Process.*, vol. ASSP-32, no. 6, pp. 1109-1121, Dec. 1984.
39. Ephraim, Y., Merhav, N., "Lower and upper bounds on the minimum mean-square error in composite source signal estimation," *Information Theory, IEEE Transactions on*, vol. 38, no. 6, pp. 1709-1724, Nov 1992.

40. Ephraim, Y., Wilpon, J.G., and Rabiner, L.R., "A Linear Predictive Front-End Processor for Speech Recognition in Noisy Environments," *Proc. IEEE ICASSP*, pp. 1324-1327, 1987.

41. Ephraim, Y., Van Trees, H.L., "A signal subspace approach for speech enhancement," *IEEE Trans. Acoust., Speech, Signal Process.*, 3(4), 251-266, 1995.

42. Ephraim Y., "Speech Enhancement Systems Using State Dependent Dynamical System Model," *IEEE Trans. on Speech and Audio Processing*, SAP–3(4): pp. 251-266, 1995.

43. Cant-Paz, E., *Efficient and Accurate Parallel Genetic Algorithms*. Springer Series: Genetic Algorithms and Evolutionary Computation, vol. 1, Springer eds., 184 p., 2000.

44. Fiscus J.G., "A Post-Processing System to Yield Reduced Word Error Rates: Recogniser Output Voting Error Reduction (ROVER)," in *Proc. IEEE ASRU Workshop*, pp. 347-352, Santa Barbara, 1997.

45. Fogel, D. B., *Evolutionary Computation: Toward a New Philosophy of Machine Intelligence*, Wiley-IEEE Press, 3rd edition, 2005.

46. Fugen, C., Holzapfel, H., Waibel, A., "Tight coupling of speech recognition and dialog management - dialog-context dependent grammar weighting for speech recognition," in *Proc. Interspeech*, pp. 169-172, 2004.

47. Gales M.J.F., and Young, S.J., "Cepstral parameter compensation for HMM recognition," *Speech communication*, vol. 12, pp. 231-239, 1993.

48. Gavsic, M., Young, S., "Effective handling of dialogue state in the hidden information state POMDP-based dialogue manager," *ACM Trans. Speech Lang. Process.*, vol. 7(3), Article 4, May 2011.

49. Goldberg, D.E., *Genetic algorithms in search, optimization and machine learning*. Addison-Wesley publishing, 1989.

50. Goldwater, S., Jurafsky, D., Manning, C.D. "Which words are hard to recognize? Prosodic, lexical, and disfluency factors that increase speech recognition error rates," *Speech Communication*, Elsevier, pp. 181-200, 2010.

51. Gong Y., "Speech Recognition in Noisy Environments: A survey," *Speech Communication*, 16, pp. 261-291, 1995.

52. Gorrell, G., Lewin, I., Rayner, M., "Adding intelligent help to mixed initiative spoken dialogue systems," in *Proc. 7th International Conference on Spoken Language Processing* (ICSLP), pp. 2065-2068, 2002.

53. Gori M., Scarselli, F., "Are Multilayer Perceptrons Adequate for Pattern Recognition and Verification?," *IEEE Trans. on Pattern Analysis and Machine Intelligence*, PAMI–20(11): pp. 1121-1132, 1998.

54. Graciarena, M., Franco, H., "Unsupervised noise model estimation for model-based robust speech recognition," in *Proc. ASRU IEEE Workshop on Automatic Speech Recognition and Understanding*, pp. 351-356, 2003.

55. Hacioglu, K., Ward, W., "Dialog-context dependent language modeling combining n-grams and stochastic context-free grammars," *Proc. IEEE ICASSP*, (ICASSP '01), pp. 537-540, vol.1, 2001.

56. Hagen, S., Morris A., "Recent advances in the multi-stream HMM/ANN hybrid approach to noise robust ASR," *Computer Speech and Language*, Elsevier, (19), pp. 3-30, 2005.

57. Haverinen H., Salmela P., Hakkinen J., Lehtokangas M., and Saarinen J., "MLP Network for Enhancement of Noisy MFCC Vectors," in *Proc. Interspeech*, pp. 2371-2374, 1999.

58. Sorensen, H.B.D., "A Cepstral Noise Reduction Multi–layer Neural Network," in *Proc. IEEE ICASSP*, pp. 933–936, 1991.

59. Henderson, J., Lemon, O., "Mixture model POMDPs for efficient handling of uncertainty in dialogue management" In Proc. 46th Annual Meeting of the Association for Computational Linguistics (ACL08), pp. 73-76, 2008.

60. Hermansky, H., "Perceptual Linear Predictive (PLP) Analysis of Speech," *J. Acoust. Soc. Am.*, 87(4), pp. 1738-1752, April, 1990.

61. Hermansky, H., & Morgan, N., "RASTA Processing of Speech," *IEEE Trans. on Audio and Speech Process.*, ASP-2(4), pp. 578-589, October, 1994.

62. Hermus, K., Wambacq, P., and Van hamme, H., "A Review of Signal Subspace Speech Enhancement and Its Application to Noise Robust Speech Recognition," *EURASIP Journal on Advances in Signal Processing*, vol. 2007, Article ID 45821, 15 pages, 2007.

63. Hernando, J., & Nadeu, C., "A comparative study of parameters and distances for noisy speech recognition," in *Proc. Interspeech*, pp. 91-94, 1991.

64. Hirsch, H. G., Pearce, D., "The AURORA Experimental Framework for the Performance Evaluations of Speech Recognition Systems under Noisy Conditions," ISCA ITRW ASR2000, Paris, September 2000.

65. Houk, C. R., Joines, J. A., Kay, M. G., "A Genetic Algorithm for function optimization: a matlab implementation," North Carolina University-NCSU-IE, technical report 95–09, 1995.

66. Cambridge University Speech Group, *The HTK Book (Version 3.4.1)*, Cambridge University Group, March 2009.

67. Hu, Y., Loizou, P., "Subjective evaluation and comparison of speech enhancement algorithms," *Speech Communication,* Elsevier, 49, pp. 588-601, 2007.

68. ITU-T, P.835, "Subjective test methodology for evaluating speech communication systems that include noise suppression algorithm," ITU-T Rec. P. 835, 2003.

69. ITU-T, P.862, "Perceptual evaluation of speech quality (PESQ), and objective method for end-to end speech quality assessment of narrowband telephone networks and speech codecs," ITU-T Rec. P. 862, 2000.

70. Jabloun, F., Champagne, B., "Incorporating the human hearing properties in the signal subspace approach for speech enhancement," *IEEE Trans. Speech Audio Process.,* vol. 11, no. 6, pp. 700-708, Nov. 2003.

71. Jakobson, R., Fant, G., & Halle, M., *Preliminaries to speech analysis: The distinctive features and their correlates*. MIT Press, Cambridge, 1963.

72. Jankowski, C., Kalyanswamy, A., Basson, S., and Spitz, J., "NTIMIT: A Phonetically Balanced, Continuous Speech, Telephone Bandwidth Speech Database," in *Proc. IEEE ICASSP*, vol. I, pp. 109–112, 1990.

73. Jelinek, F., *Statistical Methods for Speech Recognition,* MIT Press 1997.

74. Jung, Y., "Improving Robustness in Jacobian Adaptation for Noisy Speech Recognition," in *Proc. 4th IEEE tutorial and research workshop on Perception and Interactive Technologies for Speech-Based Systems,* Springer-Verlag, Berlin, Heidelberg, pp. 168-175, 2008.

75. Jurafsky, D., Martin, J. H., *An Introduction to Natural Language Processing, Computational Linguistics, and Speech Recognition,* Prentice Hall, 2nd edition, 2008.

76. Kang, S., Lee, S., Seo, J., "Dialogue Strategies to Overcome Speech Recognition Errors in Form-Filling Dialogue," in *Proc. of the 22nd International Conference on Computer Processing of Oriental Languages.* Language Technology for the Knowledge-based Economy (ICCPOL '09), Springer-Verlag, Berlin, 2009.

77. Kamper, H., Niesler, T.R., "Characterisation and simulation of telephone channels using the TIMIT and NTIMIT databases," in *Proc. of the Twentieth annual symposium of the Pattern Recognition Association of South Africa (PRASA)*, Stellenbosch, South Africa, November 2009.

78. Kim, K., Lee, C., Jung, S., and Lee, G. G., "A frame-based probabilistic framework for spoken dialog management using dialog examples," in *Proc. 9th SIGdial Workshop on Discourse and Dialogue*, pp. 120-127, 2008.

79. Klatt, D.H., "Prediction of perceived phonetic distance from critical band spectra: a first step," in *Proc. IEEE-ICASSP*, Paris, pp. 1278-1281, May 1982.

80. Kumar, A., Hansen, J.H.L., "Environment mismatch compensation using average eigenspace for speech recognition," in *Proc. Interspeech*, pp. 1277-1280, 2008.

81. Lauri, F., Illina, I., Fohr, D., Korkmazsky, F., "Using genetic algorithms for rapid speaker adaptation," in *Proc. Eurospeech*, pp. 1497-1500, 2003.

82. Ledesma R. D., "Determining the Number of Factors to Retain in EFA: an easy-to-use computer program for carrying out Parallel Analysis," *Practical Assessment*, Research & Evaluation PAR&E online, vol. 12., 2007.

83. Lee C. H., Gauvain, J.L., "Speaker Adaptation Based on MAP Estimation of HMM Parameters," in Proc. IEEE ICASSP, Minneapolis, Minnesota, pp. 558-561, 1993.
84. Legetter, C.J., Woodland, P.C., "Maximum likelihood linear regression for speaker adaptation of continuous density hidden Markov models," *Computer speech and language,* vol. 9, pp. 171-185, 1995.
85. Levin, E., Pieraccini, R., "A Stochastic Model of Human-Machine Interaction for learning dialog Strategies," *IEEE Trans. Speech, Audio Process.,* pp. 11-23, 2000.
86. Li, W.H., Yue, H., Valle-Cervantes S., Qin, S. J., "Recursive PCA for adaptive process monitoring," *Journal of Process Control,* 10(5), pp. 471-486, 2000.
87. Lim, J.S., Oppenheim, A.V., "Enhancement and bandwidth compression of noisy speech," Proceedings of IEEE, vol. 67, pp. 1586-1604, 1979.
88. Loizou, P., *Speech Enhancement Theory and Practice,* 1st Edition, CRC Press, 2007.
89. Malinowski, F.R., *Factor Analysis in Chemistry.* Wiley-Inter-science, New York, 1991.
90. Mansour, D., Juang, B.H., "A family of distorsion measures based upon projection operation for robust speech recognition," *IEEE Trans. Acoust., Speech, Signal Process.,* 37, pp. 1659-1671, 1989.
91. Mari, J.F., "HMM and Selectively Neural Networks for Connected Confusable Word Recognition," *International Conference Speech and Language Processing,* pp. 1519-1522, 1994.
92. Martin, R., "Spectral subtraction based on minimum statistics," in *Proc. of European Signal Processing Conference (EUSIPCO),* pp. 1182-1185, 1994.
93. Martin, R., "Noise Power Spectral Density Estimation Based on Optimal Smoothing and Minimum Statistics," *IEEE Trans. Speech, and Audio Process.,* 9(5), pp. 504-512, 2001.
94. Michalewicz, Z., *Genetic Algorithms + Data Structure = Evolution programs.* AI series. Springer-Verlag, New York, 1996.
95. Mokbel, C., "Online adaptation of HMMs to real-life conditions: a unified framework," *IEEE Trans. Speech, and Audio Process.,* vol.9, No 4, pp. 342-357, May 2001.
96. Moreno, P. J., Stern, R., "Sources of degradation of speech recognition in the telephone network," in *Proc. IEEE ICASSP,* vol.1, pp. 109-112, 1994.
97. Nguyen, P., Wellekens, C., Junqua, J., "Eigenspace and MLLR for speech recognition in noisy environments," in *Proc. Eurospeech,* vol. 6, Sep. 1999, pp. 2519-2522, 1999.
98. Nguyen, D., Widrow, B., "Improving the Learning Speed of Two-Layer Neural Networks by Choosing Initial Values of the Adaptative Weights," in *Proc. of IJCNN,* vol. 3, pp. 21-26, 1990.
99. Oja, E., "Neural Networks, Principal Components, and Subspaces," *Inter. Journ. of Neural Systems,* (1), pp. 61-68, 1989.
100. Paek, T., Chickering, D., "Improving command and control speech recognition: Using predictive user models for language modeling," *User Modeling and User-Adapted Interaction Journal,* 17(1), pp. 93-117, 2007.
101. Paek, T., Pieraccini, R., "Automating spoken dialogue management design using machine learning: an industry perspective," *Speech Communication* (50), Elsevier, pp. 716-729, 2008.
102. Picone, J., Signal modeling techniques in speech recognition, *Proceedings of the IEEE, 81*(9), 1215-1247, 1993.
103. Povey, D., Gales, M.J.F., Kim, D.Y. and Woodland, P.C., "MMI-MAP and MPE-MAP for acoustic model adaptation," in *Proc. of Eurospeech,* pp. 1891-1894, 2003.
104. Povey, D., *Discriminative training for large vocabulary speech recognition,* Ph.D. Dissertation, Department of Engineering, University of Cambridge, UK, 2004.
105. Qin, S. J., Dunia, R., "Determining the number of principal components for best reconstruction," in *IFAC DYCOPS'98,* Greece, June 1998.
106. Quackenbush, S., Barnwell, T., Clements, M., *Objective Measures of Speech Quality.* Englewood Cliffs, NJ: Prentice-Hall, 1988.
107. Rabiner, L.R., "A tutorial on HMM and selected applications in speech recognition," *Proceedings of IEEE,* pp. 257-286, vol. 77, No. 2, 1989.
108. Rabiner, L., Juang, B. H., *Fundamentals of Speech Recognition,* Prentice-Hall, 1993.

109. Racine, I., Detey, S., Buehler, N., Schwab, S., Zay, F., Kawaguchi, Y., "The production of French nasal vowels by advanced Japanese and Spanish learners of French: a corpus-based evaluation study," in *Proc. of New Sounds 2010 - Sixth International Symposium on the Acquisition of Second Language Speech,* pp. 367-372, 2010.

110. Rangachari, S., Loizou, P., "A noise estimation algorithm for highly nonstationary environments," *Speech Communication,* 28, pp. 220-231, 2006.

111. Rasheed, K., Hirsh., H., "Guided Crossover: A New Operator for Genetic Algorithm Based Optimization," in *Proc. of the Congress on Evolutionary Computation,* pp. 1535-1541, 1997.

112. Raut, C.K., Yu, K., Gales, M.J.F., "Adaptive training using discriminative mapping transforms," in *Proc. Interspeech,* pp. 1697-1700, 2008.

113. Rennie, S., Kristjansson, T., Olsen, P., Gopinath, R., "Dynamic noise adaptation," in *Proc. IEEE ICASSP,* vol. 1, pp. 1197-1200, 2006.

114. Rezayee A., Gazor, S., "An adaptive KLT approach for speech enhancement," *IEEE Trans. Speech, and Audio Process.,* vol. 9, no. 2, pp. 87-95, 2001.

115. Rigoll, G., "Maximum Mutual Information Neural Networks for Hybrid Connectionist-HMM Speech Recognition Systems," *IEEE Trans. Speech, and Audio Process.,* vol. 2, No. 1, Special Issue on Neural Networks for Speech Processing, pp. 175-184. 1994.

116. Rissanen, J., "Modeling by shortest data description," *Automatica,* 14, pp. 465-471, 1978.

117. Roy, N., Pineau, J., Thrun, S., "Spoken dialogue management using probabilistic reasoning" in *Proc. 38th Annual Meeting of the Association for Computational Linguistics (ACL00),* 2000.

118. Russel, R.L., Bartley, C., "The Autoregressive Backpropagation Algorithm," in *Proc. of IJCNN,* pp. 369-377, 1991.

119. Sagayama, S., Yamaguchi, Y., Takahashi, S., and Takahashi, J., "Jacobian approach to fast acoustic model adaptation," in *Proc. IEEE ICASSP,* pp. 835-838, 1997.

120. Samir, M.A., *Automatic Evaluation of Real-Time Multimedia Quality: a Neural Network Approach.* Phd. Thesis, IFSIC-IRISA, Rennes University (France), 2003.

121. Selouani S.-A., Tolba H., and O'Shaughnessy D., "Robust automatic speech recognition in low-SNR car environments by the application of a connectionist subspace-based approach to the MEL-based cepstral coefficients," in *Proc. of Eurospeech,* pp. 1577–1560, 2001.

122. Selouani, S.A., O'Shaughnessy, D., "A Hybrid HMM/Autoregressive Time-Delay Neural Network Automatic Speech Recognition System," in *Proc. European Signal Processing Conference (EUSIPCO),* paper 108, 4 pages, September 2002.

123. Selouani S.-A., and O'Shaughnessy D., "Robustness of speech recognition using genetic algorithms and a Mel-cepstral subspace approach," in *Proc. IEEE ICASSP,* vol. I, pp. 201–204, 2004.

124. Selouani, S.A., O'Shaughnessy, D., "Speaker adaptation using evolutionary-based linear transform," in *Proc. of International Conference on Spoken Language Processing,* pp. 1109-1112, Pittsburgh, November 2006.

125. Selouani, S.A., "Using Robust and Flexible Speech Recognition Capabilities in Clean to Noisy Mobile Environments," *Advances in Speech Recognition: Mobile environments, Call Centers and Clinics,* Neustein, Amy (Ed.), pp. 91-112, 2010.

126. Shah, S.A.A., Ul Asar, A., Shah, S.W., "Interactive Voice Response with Pattern Recognition Based on Artificial Neural Network Approach," *International IEEE conference on Emerging Technologies,* pp. 249-252, 2007.

127. Shukla, A., Tiwari, R., Kala, R., *Real Life Applications of Soft Computing.* CRC Press, ISBN: 1439822875, 686 pages, 2010.

128. Singh, S., Litman, D., Kearns, M., Walker, M., "Dialogue Management with Reinforcement Leaning: Experiments with the NJFun System." *Optimizing Journal of Artificial Intelligence,* vol. 16, pp. 105-133, 2002.

129. Sivanandam, S.N., Deepa, S.N., *Introduction to Genetic Algorithms,* Springer eds., 1st edition, 2007.

130. Sohn, J., Sung, W., "A voice activity detector employing soft decision based noise spectrum adaptation," in *Proc. IEEE ICASSP,* pp. 365-368, 1998.

131. Spalanzani, A., Selouani, S.A., Kabre, H., "Evolutionary algorithms for optimizing speech data projection," *Genetic and Evolutionary Computation Conference*, Orlando, pp. 1799, 1999.
132. Temby, L., Vamplew, P., Berry, A., "Accelerating Real-Valued Genetic Algorithms Using Mutation-With-Momentum," *Springer Lecture Notes in Computer Science series*, Australian joint conference on artificial intelligence, vol. 3809, pp. 1108-1111, Sidney, Australia, 2005.
133. Fisher, W.M., Dodington, G.R., Goudie-Marshall, K.M., "The DARPA Speech Recognition Research Database: Specification and Status," in *Proc. DARPA Workshop on Speech Recognition*, pp. 93–99, 1986.
134. Tohkura, Y., "A weighted cepstral distance measure for speech recognition," *IEEE Trans. Acoust., Speech, Signal Process.*, vol. ASSP-35, pp. 1414-1422, Oct.1987.
135. Tolba, H., Selouani, S.A., O'Shaughnessy, D., "Auditory-based acoustic distinctive features and spectral cues for automatic speech recognition using a multi-stream paradigm," in *Proc. IEEE ICASSP (ICASSP'2002)*, pp. 837-840, 2002.
136. Tufts, D.W., Kumaresan, R., Kirsteins, I., "Data adaptive signal estimation by singular value decomposition of a data matrix," *Proceedings of the IEEE*, vol. 70, no. 6, pp. 684-685, 1982.
137. Uebel, L.F., Woodland, P.C., "Discriminative linear transforms for speaker adaptation," in *Proc. of ISCA ITRW Adaptation Methods for Automatic Speech Recognition*. Sophia-Antipolis, France, pp. 61-63, 2001.
138. Uemura, Y., Takahashi, Y., Saruwatari, H., Shikano, K., Kondo, K., "Musical noise generation analysis for noise reduction methods based on spectral subtraction and MMSE STSA estimation," in *Proc. IEEE ICASSP.* pp. 4433-4436, 2009.
139. Visweswariah, K., Printz, H., (2001): "Language models conditioned on dialog state," in *Proc. Eurospeech*, pp. 251-254, 2001.
140. Waibel, A., Hanazawa, T., Hinton, G., Shikano, K., Lang, K., "Phoneme Recognition Using Time Delay Neural Networks," *IEEE Trans. Acoust., Speech, Signal Process.*, 37, pp. 328–339, 1989.
141. Wang, S., Sekey A., and Gersho A., "An objective measure for predicting subjective quality of coders," *IEEE Journal on Selected Areas Commun.*, (10), pp. 819-829, 1992.
142. Wang, Z., Schultz, T., and Waibel, A., "Comparison of Acoustic Model Adaptation Techniques on Non-Native Speech," in *Proc. IEEE ICASSP*, pp. 540–543, 2003.
143. Wang, L., *Discriminative linear transforms for adaptation and adaptive training*, Ph.D. Dissertation, Department of Engineering, University of Cambridge, UK, 2006.
144. Wang, L., and Woodland, P.C., "MPE-based discriminative linear transforms for speaker adaptation," Computer Speech and Language, vol. 22(3), pp. 256-272, 2008.
145. Wang, L., Woodland, P.C., "MPE-based discriminative linear transform for speaker adaptation," in *Proc. IEEE ICASSP*, vol. I, pp. 321–324, 2004.
146. Williams, J., and Young, S., "Partially observable Markov decision processes for spoken dialog systems," Computer Speech and Language, Elsevier, (21), pp. 393-422, 2007.
147. Woodland, P.C., and Povey, D., "Large scale discriminative training of hidden markov models for speech recognition," Computer Speech and Language, 16, pp. 25-47, 2002.
148. Yang, S., Bosch, L.T., Boves, L., "Hybrid HMM/BLSTM-RNN for robust speech recognition," in *Proc. 13th international conference on Text, speech and dialogue (TSD'10)*, Springer-Verlag, Berlin, Heidelberg, pp. 400-407, 2010.
149. Yang, W., Dixon, M., Yantorno, R., "A modified bark spectral distortion measure which uses noise masking threshold," in *Proc. IEEE Speech Coding Workshop*, pp. 55-56, Pocono Manor, 1997.
150. Yang, W., Benbouchta, M., Yantorno, R., "Performance of a modified bark spectral distortion measure as an objective speech quality measure," in *Proc. IEEE ICASSP*, pp. 541-544, Seattle, 1998.
151. Young, S., Gasic, M., Keizer, S., Mairesse, F., Schatzmann, J., Thomson, B., and Yu, K., "The Hidden Information State Model: a practical framework for POMDP-based spoken dialogue management," Computer Speech and Language, 24(2), pp. 150-174, 2010.

152. Yu, K., Gales, M., Woodland, P.C., "Unsupervised Adaptation With Discriminative Mapping Transforms," *IEEE Trans. Audio., Speech, and Language Process.*, vol.17, no.4, pp. 714-723, May 2009.
153. Zadeh, L. A., "Fuzzy Logic, Neural Networks, and Soft Computing," *Communications of the ACM*, vol. 37 No. 3, pp. 77-84, March 1994.